K. S. Spiegler

Principles of Energetics

Based on
Applications de la thermodynamique du
non-équilibre
by P. Chartier, M. Gross, and K. S. Spiegler

With 21 Figures

Springer-Verlag
Berlin Heidelberg New York Tokyo 1983

Professor Emeritus K. S. SPIEGLER
University of California, Berkeley
College of Engineering
Berkeley, CA 94720/USA

ISBN-13: 978-3-642-95436-8   e-ISBN-13: 978-3-642-95434-4
DOI: 10.1007/ 978-3-642-95434-4

Library of Congress Cataloging in Publication Data. Spiegler, K. S. Principles of energetics.
Includes bibliographical references. 1. Power (Mechanics) 2. Thermodynamics. 3. Force and
energy. I. Title. TJ163.9.S69 1983   621.042   83-12431 ISBN 0-387-12441-1 (U.S.)

Dedicated to my wife Annie, whose under-
standing and patience made this work possible

Preface

The purpose of this book is to lay the groundwork for the analysis and the design of processes with a view to energetic efficiency. *Energetics* is used in the original sense of the engineer W. J. M. Rankine (Proc. Roy. Philosoph. Soc. of Glasgow III, 381 [1955]) and the physical chemist J. N. Brønsted (*Principles and Problems in Energetics,* Interscience, New York, 1955), i.e., the macroscopic description of the flows of different forms of energy, and the general laws governing the mutual transformations of these flows.

The prerequisite for the use of the book is a conventional course in *equilibrium thermodynamics* as usually taken in the junior (third) college year. The author believes that while knowledge about equilibria is essential, most engineers and many scientists are mostly interested in systems in which equilibrium has not yet been reached. In such systems, flow phenomena such as heat, mass and electricity transfer, as well as chemical reactions, can take place, and it is important to know the driving forces and laws governing the interactions of these flows.

The key element in this text is the concept of *exergy*—often called available (or utilizable) energy—which is mechanical or electrical or any other energy which, in principle, can be completely converted into useful mechanical or electrical work in the presence of the given environment. After restating the principles of thermodynamics, adapted for the use of exergy as the central concept of energetics, the laws governing the relations between (a) flows and generalized driving forces, and (b) simultaneous flows are discussed. These laws are then applied to a number of coupled flow processes. Only "linear" processes are discussed here, i.e., situations in which flow (e.g., electric current) and conjugated generalized driving force (e.g., voltage) are proportional to each other.

I have tried to bring together methods from the fields of (a) engineering thermodynamics, which places the *rate of exergy dissipation* in the center of modern energetic analysis (second-law analysis), and (b) the thermodynamics of irreversible processes, which places the *rate of entropy creation* in the focus of its treatises. In effect, these concepts represent two sides of the same coin, since they are related by a simple formula, the Maxwell-Gouy-Stodola equation, which also introduces the temperature of the *environment*, and is discussed in Chapter II. In addition, some *thermoeconomic fundamentals* are presented in this context.

This text is for a one-quarter or one-semester course presented in the senior (fourth) year of undergraduate studies in engineering, or in the first year of graduate studies for chemical and mechanical engineers. Since biophysicists and some civil engineers (e.g., soil mechanicists) and physical chemists have participated in my course on this subject, I hope that the material presented here, as a supplement to traditional courses of equilibrium thermodynamics, is of genuinely interdisciplinary nature, and that it will help in the understanding and design of energy-conversion processes.

The solution of the *problems* should enhance the comprehension of the main text. In the last two decades, I have used most of these problems in my courses, and many of the numerical data are from research accounts.

The lists of selected literature at the end of the chapters are to provide the reader with additional study materials to complement the main text of the chapters, where several, but not all references, are specifically mentioned. These lists contain books and articles which have helped me in the presentation of the subjects; I am indebted to the authors for their teachings. These bibliographies are not complete compilations of the pertinent literature, however. Many are reviews which are to assist in the search for details. In the hope of helping the reader in this task, I have added brief remarks where the titles of the books or articles do not reveal the specific points to which I wish to draw attention.

July 1983   K. S. SPIEGLER

Contents

Contents   XI

Chapter I. Fundamental Concepts

A. Introduction

The purpose of this chapter is the presentation of definitions which are essential for the comprehension of this volume, and of the postulates (Sect. I. C) which restate the laws of thermodynamics adapted for the exergy-oriented approach of this book. Not all derivations and laws of equilibrium thermodynamics are reviewed here. For details, the reader is referred to many excellent texts of equilibrium thermodynamics in the literature, some of which are listed under Selected Literature at the end of this chapter.

In this scheme of definitions, *work* is the basic concept. The concept of temperature is also used. Note that Kelvin's scale of *absolute temperature* is based on the concepts of (a) work and (b) heat flow [expressed by Eq. (I-6) in terms of work and energy, which, in turn is defined from work as shown in this chapter]. Therefore, temperature does not represent an independent concept in this scheme.

B. Definitions

Work

According to Maxwell's definition (1878, p 101), work is the "act of producing a change of configuration in a system in opposition to a force which resists that change." Note that this definition specifically states that the force has to *oppose* the change. Thus, not every integral of (force × element of length) is thermodynamic work. Lifting a weight is work because the motion is done against resistance; dropping the weight without control only causes sound, deformation and heat flow (*dissipation*).

The production of work is the lifting of weights, or compression of a spring, or activation of an electric motor which charges a condenser, or the turning of a pump which lifts water from a well, or the performance of any equivalent action. The amount of work equals the weight times the distance of the lift. (An equivalent action can produce any of these results without causing additional permanent changes; hence the production of heat is not equivalent in this sense, because the reconversion of heat to perform the work necessary to lift a weight, for example, necessitates a permanent change in another system, viz., a body of lower temperature to which some heat is conducted.[1]) The action of a force along a distance is a necessary, but not a sufficient condition for the thermodynamic definition of work production.

System

A *system* consists of the contents of a definite region in space. To enable us to define *macroscopic* properties of the material composing the system, it is necessary that the volume and the masses contained in it be sufficiently large. The limits of each system have to be well defined.

Extensive and Intensive Functions

Extensive functions are those whose values increase in direct proportion to the size of the system. Mass or net electrical charge are such functions. On the other hand, intensive functions are independent of the size of the system. Density, temperature and dielectric constant are intensive functions. (Note, however, that these terms have to be used judiciously. For instance, the terminal voltage of a group of batteries is independent of the number of batteries when they are connected in parallel, but proportional to the number of batteries when they are connected in series.)

Thermodynamic Process

Consider a system at two times, t_1 and t_2. A thermodynamic process has taken place between these two times if at least one of the properties at t_1 is different at any time t_2. This definition includes thermodynamic *cycles* because, although the initial and final states are equal, the system is different from the initial state (t_1) at an intermediate time, t_2.

State Function

Such a function, Y, is defined as a *property* which unequivocally defines the state of a system (with respect to this property) independent of the way by which it arrived at its state.

---

1 See the discussion of the *Carnot cycle* in Appendix I, or in general textbooks of thermodynamics, e.g., Hatsopoulos and Keenan 1965.

Suppose a system undergoes a change from state **A** to state **B** (with or without interaction with other systems). If Y is a state function, the change of Y is equal to $Y_B - Y_A$ ($\equiv \Delta Y$), and is independent of the path **A** → **B**.

Some functions do not satisfy this criterion, e. g., the work or the heat produced in the transformation **A** → **B**. Therefore, work and heat are not state functions.

Reservoir

Since the ability of many systems to produce work depends on the nature of the *environment* with which the system interacts, it is useful to introduce an idealized environment concept at an early stage of energetic considerations. The environment is represented by a reservoir (or medium) which is a system which cannot furnish work by itself, even with the help of non-changing external devices (e. g., catalysts or inert electrodes).

A reservoir must be sufficiently *large* so that the changes of temperature, pressure, and all other intensive properties, caused by interaction with the systems considered are negligible.

Since one can utilize temperature and/or pressure differences for the production of work, it follows that in any reservoir, entirely isolated from other bodies, temperature and pressure are uniform.

A body of still air or the water of a calm lake would approach the properties of reservoirs, except for hydrostatic pressure differences, but in nature not even these reservoirs are entirely perfect.

Some functions used for the energetic characterization of systems depend not only on the state of the system, but also on the environment. They are not state functions in the strict sense of the definition, but are called *potential functions.*

Exergy

The *exergy* of a system, with respect to a reservoir of given temperature and pressure, is the maximum useful work which the system can produce without causing permanent changes in any other system (except the reservoir). *Useful* work is the work produced by the system − *except for the work due to volume changes of the system against the given reservoir pressure.*

The *reference reservoir* chosen here should be sufficiently large, so that its interaction with the system causes only infinitesimal changes of temperature, pressure and all other intensive properties of the reservoir.

The term exergy is a relatively recent word (Rant 1956, 1957) for a much older concept which has been discussed before and after the coining of this name under various other names and with equal, or slightly different meanings, e. g., *energie utilisable* (Gouy 1889, Van Lerberghe and Glansdorff 1932); *available energy* (Obert 1960); *availability* (Hatsopoulos and Keenan 1965, Denbigh 1971), and *essergy* (R. B. Evans 1969). The history of this concept is discussed in an appendix by W. Fratzscher to the text by Wukalowitsch and Nowikow (1962) and by Haywood (1974). Detailed discussions about this parameter, including its sign, are in Chapter II of this text. From them it can be concluded that if the reference reservoir is at lower temperature and pressure than the system, and if work is produced only as a result of the temperature differences (Carnot engine) and pressure differences, the exergy of the system is *positive*: the system can *produce* work.

In the special case of system and reservoir having identical temperatures and pressures, the exergy is equal to the difference between the free energy (Gibbs function, G) of the system (a) in its original state, and (b) in its final state, i. e., when no more work can result from interaction of the system with the reservoir. In other words, the Gibbs function of the system is equal to the exergy of the system (provided the state of the system when in equilibrium with the reservoir is taken as the reference state).

Total Energy

The *total energy*, E, of a system contained in a rigid enclosure is its maximal exergy with respect to a reservoir of lower temperature, i. e., its exergy with respect to a reservoir which by interaction with the system can produce more work than *any other reservoir of lower temperature* (see Postulate 2, Sect. I. C). The temperature of this reservoir is chosen as reference temperature (0 K).

The *total energy* is a state function, i. e., independent of the path chosen from original to reference state, and will often be simply termed *energy*, omitting the term *total*.

Most, but not all, problems arising in thermodynamics require only a knowledge of energy *differences* of a system between two states rather than of absolute energy values. This is also true of the other state functions. Therefore, one may subtract an arbitrary constant from both initial and final value of the state function, and yet obtain the correct value for the difference of the state function in the two states.

Internal Energy

The *internal energy*, U, of a system is equal to the (total) energy of the system, when the system is in the following state:
a) the kinetic energy of the system is zero;
b) no external fields (gravity, electric, etc.) act on the system;

c) the system is not subjected to *elastic tensions*, and the contribution of *surface forces* to the specific energy of the system is negligible.

When not all of these conditions are satisfied, the *total* energy of the system is different from its *internal* energy.

Conditions b) and c) define a simple system (Hatsopoulos and Keenan 1965, p 64). If, in addition, condition a) is satisfied, the system is a *simple system at rest*. When such a system can no longer produce work by interaction with a given reservoir, it is completely relaxed with respect to this particular reservoir.

For a simple system at rest, the internal energy equals the total energy. Thus, the internal energy[2] is simply the contribution of the atomic and/or molecular motions — when unperturbed by external constraints — to the total energy.[3] It is the energy listed in thermodynamic tables.

Heat

When a system undergoes a transformation, the heat, Q, which flows into the system across its boundary is, by definition, the algebraic sum of the energy increase of the system and the work produced which leaves the system. This is the thermodynamic definition of the flow of heat, Q_{therm}. One of the important empirical laws of macroscopic thermodynamics (introduced as Postulate 3, Sect. I.C) is the equality of the heat flow defined in this manner with the heat flow measured calorimetrically, Q_{cal}. Imagine a well-stirred water — ice mixture enclosed in a vessel made of metal (or another good conductor of heat), and otherwise isolated, the metallic enclosure being brought into contact with a reservoir. One unit of caloric heat (1 J = 1 W s) has entered the mixture from the reservoir when 0.00299 g of ice melt at atmospheric pressure. Other things being equal, the faster the rate of melting, the higher the temperature of the reservoir.[4]

Enthalpy

The *total enthalpy*, H, of a system of volume V and pressure p is:

$$H \equiv E + pV. \tag{I-1}$$

---

2 In some texts, the term internal energy and the symbol U are used to designate the *total* energy (e.g., Katchalsky and Curran 1965).

3 If *nuclear* properties are to be taken into account, due regard must be paid to the inclusion of nuclear terms in the definition of the reference reservoir.

4 If ice does not melt, but water freezes instead, then the heat flow is from the vessel to the reservoir, i.e., in the opposite direction. The formation of 0.00299 g of ice corresponds to heat flow of one joule. The faster the rate of freezing, the lower the temperature of the reservoir.

If the system consists of different regions (subsystems) in which different pressures prevail, the enthalpy of the system as a whole is the sum of the enthalpies of the different regions. The enthalpy of each region can be calculated if the region has a well-defined uniform pressure, which is often the case for macroscopic regions, e.g., a mass of ice immersed in water of the same temperature at uniform pressure.

For a simple system at rest, the *total* energy, E, can be replaced by the *internal* energy, U, yielding the tabulated enthalpy, H_u, which is the enthalpy value usually found in thermodynamic tables, $H_u = U + pV$ (note that tables list the enthalpy for systems of unit mass). Thus, for a simple system *at rest*, consisting of unit mass of a pure substance, the tabulated enthalpy and the total enthalpy are identical.

Sign Conventions About Heat and Work

While it is useful to consider all quantities entering a system as positive, and all those leaving the system as negative, this sign convention is not adopted in this book. For the most part, the traditional sign conventions for closed systems have been used, viz.:

Work, W, which is *produced* by a system (e.g., the work output of an engine) is counted *positive*;

Heat, Q, *entering* a system is counted *positive*, while heat leaving a system is counted negative.

Reversible Process

This term designates a process taking place under reversible conditions, which means that at every stage an appropriate infinitesimal change of the driving force (i.e., a change which is as small as one wishes) reverses the direction of the process. In the limit, the process is infinitely slow and may be considered as a succession of steps in which the difference between driving forces and resisting forces is infinitesimal.

Thus the interdiffusion, through a valve, of two gases originally held separately in two vessels is not a reversible process, nor is the generation of heat by friction.

As an illustration of a reversible process, consider a system consisting of a 0.127 molar solution of hydrochloric acid[5] at 25 °C, a standard hydrogen and a standard chlorine electrode.[6] Under these conditions, the electromotive force of

---

5 The mean ionic activity of H^+ and Cl^- in this solution is 0.1.

6 These electrodes consist of platinized platinum wires surrounded by atmospheres of hydrogen (H_2, 1 atm) and chlorine (Cl_2, 1 atm), respectively.

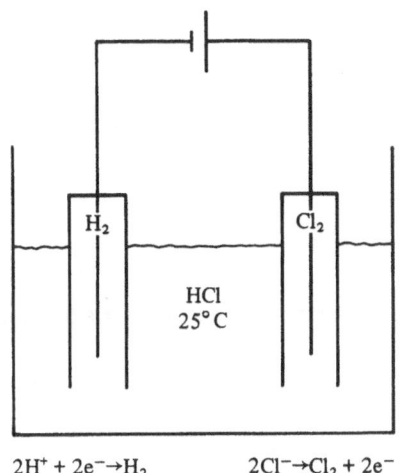

Fig. I-1. Reversibility. Electrolysis of a solution of hydrochloric acid is shown. The required voltage is at least 1.477. If the voltage is permitted to drop below this value, the cell acts as a battery, and all reactions are reversed. If performed near this critical voltage, the process is said to be *reversible*

Electrolysis	$2H^+ + 2e^- \rightarrow H_2$	$2Cl^- \rightarrow Cl_2 + 2e^-$
Fuel cell	$H_2 \rightarrow 2H^+ + 2e^-$	$Cl_2 + 2e^- \rightarrow 2Cl^-$

Total reaction

Electrolysis $2HCl \rightarrow H_2 + Cl_2$
Fuel cell $H_2 + Cl_2 \rightarrow 2HCl$

the cell, i.e., the electric potential difference which can be measured between the protruding ends of the platinum wires, is 1.477 V (Fig. I-1).

Now perform two experiments, viz.:

1. Apply to the electrodes of this cell an electric potential difference of 1.480 V (or any other voltage slightly larger than 1.477). This can be done by connecting the electrodes to a suitable battery or other direct-current source. This will cause electrolysis in the cell, resulting in the *production* of hydrogen (H_2) and chlorine (Cl_2) gas from the hydrochloric acid (HCl).

2. Connect the electrodes of the cell by means of an electric conductor. The voltage of the cell will immediately drop, as an electric current flows in the direction opposite to the current in experiment 1. If a conductor of high electric resistance is chosen, the voltage drops only slightly, say to 1.474. Now the system acts as a battery, *consuming* hydrogen and chlorine gas, and producing hydrochloric acid. In fact, in this mode, the cell is often called a fuel cell.

Thus, if the voltage is maintained close to 1.477, the cell operates under *reversible* conditions, because the slightest appropriate change of the voltage can entirely invert the direction of the cell process, i.e., both the directions of the electro-chemical reactions and of the electric current. On the other hand, if — in experiment 1 — electrolysis were performed faster by applying 2.5 V, the conditions would be irreversible, because the process could not be reversed by the lowering of the voltage by a mere few mV. Similarly, if in experiment 2 the cell were discharged through a much lower resistance, thus reducing the cell voltage to, say, 1.2 V (and yielding a higher electric current), the conditions would be considered *irreversible* for similar reasons.

Irreversible Process

This is a process which does not satisfy the reversibility criteria discussed in the preceding paragraph.

Isothermal Process

A process is isothermal if it proceeds at uniform and constant temperature.

Monothermal Process

A system undergoes a monothermal process if heat can be transferred to or from a single reservoir only, which does not necessarily have to be at the same temperature as the system.[7]

When a system changes from state **A** to **B** by a monothermal process, all *reversible* paths produce (or consume) the same amount of work (a proof of this statement is presented in Sect. I. C). The absolute value of this work produced (or consumed) is larger than for any irreversible process leading from state **A** to state **B**. These processes often involve heat transfer to or from the reservoir. A discussion on how to envisage *reversible heat transfer* between two bodies at different temperatures is presented in Sect. II.B.2.

Thus, the term *maximum* work used in the definition and discussion of exergy (Chaps. I and II) always refers to any *reversible* path involving the system and the reservoir.

Isolated System

By definition, no work, heat or mass transfer can take place into or from such a system.

Adiabatic System

By definition, no heat can be transferred across the boundary of such a system.

---

7 Energetics often deals also with dithermal processes (e. g., the production of power by a steam engine or, conceptually, by a Carnot engine) and polythermal processes.

Closed System

No mass transfer can take place across the boundary of a closed system. Heat and/or work flows may occur across these boundaries. By definition, isolated systems are closed systems.

Open System

An "open system" is a defined region of space (control volume). Mass, heat and/or work flow can take place across its boundary. Such control volumes are used extensively in establishing mass and energy balances for elements of plants, engines, mass-transfer devices, etc. The control volume is not necessarily fixed in space, but can be defined as moving (e. g., in the analysis of rocket propulsion).

Global System

Non-isolated systems can interact with other systems, including reservoirs. In analyses, it is often convenient to consider the whole assembly of interacting systems, and define the whole as the *global* system.

For instance, the steam turbine of a power plant is a system which interacts with two other systems, viz., the boiler and the condenser. Turbine, boiler and condenser may be taken together as a global system.

Entropy

The entropy of a system in state **A** is defined by imaging the system to be taken from its standard state, **0**, to **A** by a reversible process; in each infinitesimal stage of this process, the heat flow, dQ_{rev} to or from the system is monitored and divided by the (absolute) temperature, T, at which the heat flow takes place. Then all the parameters, dQ_{rev}/T, are added:

$$S_A \equiv \int_0^A \frac{dQ_{rev}}{T} . \tag{I-2}$$

Classical thermodynamics demonstrates that the same value for the entropy is obtained, no matter what kind of reversible process $0 \to A$ is chosen (see, for instance, Hatsopoulos and Keenan 1965, p 154; or Fabry 1930, p 84). The standard state, **0**, must be well defined; for instance, in thermodynamic tables for homogeneous substances the rest state at 25 °C, 1 atm, is often taken as the origin of the entropy scale for the respective substance. For complex systems, the zero point is sometimes the state approached in a thought experiment, in which all components are separated, and solidified to perfect crystals while cooling them to 0 K.

In the transformation of the system from state **0** to **A**, described by Eq. (I-2), work and/or heat may be transferred across the system boundary when required, provided all transfers take place *reversibly*. Heat can flow *reversibly* from a reservoir into a system when the reservoir temperature is only infinitesimally higher than the system temperature at the point of heat transfer. (If the temperature difference were not *infinitesimally* small, the direction of the heat flow could not be reversed by an extremely small change of the reservoir temperature; hence, by the definition of reversibility given in this chapter, the heat transfer would take place under irreversible conditions.[8])

Moreover, when a *reversible* process takes place within a global system, the entropy of this system is conserved. This means that the increase of entropy which might occur in one subsystem is compensated by a numerically equal decrease of entropy in other subsystem(s) of the global system.

Equation (I-2) implies that when a closed system of uniform temperature, T, changes from state **A** to another state in the vicinity of **A**, the change of entropy is

$$dS = \frac{dQ_{rev}}{T}.$$
(I-3)

If the change is irreversible, a non-compensated *increase* of the entropy of the global system takes place:

$$d_i S > 0 \quad \text{(irreversible process)}.$$
(I-4)

(The subscript, i, next to the differential operator, d, indicates that the entropy increase is due to an internal process in the system.) This principle of entropy increase due to internal changes in global systems (Clausius principle) is proven in texts of classical thermodynamics (see, for instance, Katchalsky and Curran 1965, p 43, Prigogine and Defay 1950, p 36).

C. Postulates

The fundamental laws of thermodynamics have been formulated in different ways.[9] Though different in their approach and logical structure, many of these

---

8 A thought experiment for reversible heat transfer across a finite temperature difference is described in Sect. II.B.2.

9 See, for instance, the review of Keenan and Shapiro (1947) or the introduction to Baehr's text (1966).

formulations are indeed equivalent to each other (see, for instance, the 22 formulations of the second law listed by Campbell 1970). An attempt has even been made to combine the first and second laws into a single one (Hatsopoulos and Keenan 1965, p 367). In many classical formulations, the basic concept is *energy*, although it has been stated that

> "As science progressed it has been necessary to invent other forms of energy, and indeed an unfriendly critic might claim, with some reason, that the law of conservation of energy is true because we make it true by assuming the existence of forms of energy for which there is no other justification than the desire to retain energy as a conservative quantity." (Lewis and Randall 1923, p 29 of 1961 ed.)

For many engineers, however, and also for a good number of scientists, the fundamental concept is the work, W, obtained from a given process, or often the *useful work* (i. e., the total work minus the work produced against the pressure of the reservoir, e. g., against the atmospheric pressure). The maximum useful work which a system in state **A** can produce until it reaches final equilibrium with a reservoir of given temperature and pressure is the *exergy*, Λ_A, of the system with respect to the given reservoir.

This book is based on two postulates founded on the concept of *work* and an empirical law about *heat* flow (Postulate 3). All three are related to the classical well-known laws of thermodynamics, which have been reformulated and rearranged so as to consistently place the exergy concept into the focus of this presentation.

Postulate 1. If a system cannot produce work (without the help of other systems in which a permanent change takes place), then it cannot change spontaneously, and vice versa. [10]

Such a system is in the state of *thermodynamic equilibrium*. By definition, a *reservoir* is always in equilibrium.

It is seen that this postulate defines the state of thermodynamic equilibrium. [11] This term will be used for *stable* equilibrium only. In mechanics, a ball resting on

---

10 J. A. V. Butler has formulated the second law of thermodynamics as follows:
  "Spontaneous processes (i. e., processes which may occur of their own accord) are those which when carried out under proper conditions, can be made to do work. If carried out reversibly, they will yield a maximum amount of work; in the natural irreversible way the maximum work is never obtained." (Quoted by Campbell 1970, p 1089).

11 In this text, the term *thermodynamic equilibrium* refers only to complete equilibrium, i. e., when the condition of Postulate 1 is completely satisfied. For instance, the state of rest of two aqueous solutions of common salt of different concentrations, separated by an ideally semipermeable membrane (described in Chap. VIII) and held under a pressure difference equal to the osmotic pressure difference between the solutions, is not in complete thermodynamic equilibrium. By replacing the membrane by a non-ideal one, and inserting silver-silver chloride electrodes in the solutions, electric power can be

either a valley floor or a mountain peak is considered to be in equilibrium (stable and labile equilibrium, respectively). In the second case, the slightest displacement from equilibrium causes the ball to roll downhill. This effect could be used to produce work. Hence, thermodynamic equilibrium is not a labile equilibrium (Principle of Le Chatelier); otherwise, even thermal motions of molecules could precipitate major spontaneous changes in systems at equilibrium.

One can prove, by use of this first principle, that *the exergy of any system (with respect to a given reservoir) depends only on the state of the system and not how it arrived at this state*, i.e., that exergy is a *potential* function. In order to prove this point, consider a monothermal, reversible transformation of the system from its initial state, $A(T_A, p_A, V_A)$ to a state $B(T_0, p_0, V_0)$, in which the system is in equilibrium with the reservoir. This transformation proceeds by a sequence of events, α. Next, the system returns to its original state by a different route, β.[12] During all processes, the system can interact only with the single reservoir at T_0, p_0 since the total process is stated to be *monothermal*. (In Chap. II, a method for effecting heating or cooling monothermally and *reversibly* is described − see Fig. II-1.) If the process $(A \rightarrow B)_\alpha$ produced more work than the return route $(B \rightarrow A)_\beta$ consumed, then the total cycle would produce work, although the system returns to its original state and interacts with the single reservoir. In other words, one could extract work from a single reservoir without permanently changing any other system. But this is impossible according to the first principle stated here (or the classical second law, which is equivalent to it). Hence, the sequence $(A \rightarrow B)_\beta$ cannot produce more work than the route $(B \rightarrow A)_\alpha$ consumes.

Since the whole cycle is reversible, one can reverse each stage of the cycle and the *directions* of work and heat flows, leaving the absolute values of these flows essentially the same. In this case, the system would change from state A to B via route β, and return via route α. Again $(A \rightarrow B)_\beta$ cannot produce more work than $(B \rightarrow A)_\alpha$ consumes, which is numerically essentially equal to $(A \rightarrow B)_\alpha$. We conclude that the two reversible routes $(A \rightarrow B)_\alpha$ and $(A \rightarrow B)_\beta$ produce the same

---

obtained without causing net changes in the composition of these auxiliary systems. (The electrode in the more concentrated solution loses silver and gains silver chloride, while exactly the opposite occurs at the other electrode, so that no *net* change in the composition of the environment as a whole takes place.) Therefore, this situation, *which is characterized by the absence of a net driving force for the transport of water, but not for salt*, does not conform to the definition of (complete) thermodynamic equilibrium, formulated in Postulate 1.

12 For instance, if the system is a gas at T_A, p_A, contained in a cylinder with piston $(T_A \gg T_0, p_A \gg p_0, V_A \ll V_0)$ route $(A \rightarrow B)_\alpha$ might be the following:

  $1._\alpha$ adiabatic expansion to volume V_0,
  $2._\alpha$ cooling at constant volume to T_0.

The return route $(B \rightarrow A)_\beta$ might be:

  $1._\beta$ isothermal compression at T_0 from V_0 to V_A,
  $2._\beta$ heating at constant volume from T_0 to T_A.

amount of work, which is the maximum work which can be produced by the monothermal transformation **A** → **B**. (If an *irreversible* monothermal process could produce *more* work, the same kind of argument shows that the first principle would be breached.) This is also true for the maximum amount of *useful* work, because the pressure within the system in state **B** is equal to the rervoir pressure. This amount of useful work is, by definition, the *exergy*, $\Lambda_\mathbf{A}$, of the system in state **A**, with respect to the given single reservoir at T_0 and p_0.

Postulate 2. When measuring the exergies of a system contained in a rigid enclosure with respect to different reservoirs, starting with a reservoir of the same temperature as the system, and consecutively decreasing reservoir temperature, a reservoir is assymptotically approached which gives a higher exergy than any other.[13]

While this ideal reservoir cannot be produced in practice, one can − in principle, at least − approach a reservoir with properties as close to the ideal as one wishes.

By definition, this maximal exergy of the system is its *total energy*, E.[14] As with any exergy with respect to a reservoir of given properties, E depends only on the properties of the system. The temperature of this reservoir is − by definition − the origin of the *absolute temperature scale* (0 K). Because the definition of energy relates to a specific reservoir, permitting no arbitrary choice of reservoir temperature, energy is a genuine state function.

Consider a system which changes from state $\mathbf{A}(T_\mathbf{A}, p_\mathbf{A})$ to $\mathbf{B}(T_\mathbf{B}, p_\mathbf{B})$. The total energy change for the change **A** → **B** is

$$E_\mathbf{B} - E_\mathbf{A} = \Lambda_{\max, \mathbf{B}} - \Lambda_{\max, \mathbf{A}} \, . \tag{I-5}$$

Since $\Lambda_{\max, \mathbf{A}}$ and $\Lambda_{\max, \mathbf{B}}$ depend only on the states **A** and **B** of the system, respectively (the reservoir having been specified), the energy change of the system is independent of the path of its transition from state **A** to state **B**.

Postulate 3. The heat flowing into a system, measured by calorimetry, equals the algebraic sum of the energy change of the system and the work output, irrespective of the physical nature of the work.

$$(Q_{\mathbf{A} \to \mathbf{B}})_{\text{cal}} = (Q_{\mathbf{A} \to \mathbf{B}})_{\text{th}} \equiv (E_\mathbf{B} - E_\mathbf{A}) + W_{\mathbf{A} \to \mathbf{B}} = \Delta E + W_{\mathbf{A} \to \mathbf{B}} \, . \tag{I-6}$$

There are two points made here, namely (1) the equality of *caloric* heat, Q_{cal}, as defined in elementary treatises on heat flow, and of *thermodynamic* heat, Q_{th}, defined in Eq. (I-6), as shown by many experiments described in texts of equi-

---

13 J. C. Maxwell (1888, p 185) stated that "if we possessed a perfect reversible engine and a refrigerator at the absolute zero of temperature, we might convert the whole of the heat which escapes from the body into mechanical work." See also Loebl (1960).

14 In the discussion of Postulate 3, it will be shown explicitly that in an isolated system the total energy does not change as a result of internal changes.

librium thermodynamics (e. g., the early work of Rumford and of Joule), and (2) the fact that the sum $\Delta E + W$ is the same, irrespective of the physical nature of the work, e. g., lifting a weight, compressing a steel spring, charging a condenser, etc.

In Eq. (I-6), all parameters, including Q_{cal}, are expressed in W s. The conversion factor to calories, 0.2389 cal $(W s)^{-1}$, was already determined in the classical experiments in the early 19th century.

In general, the work, $W_{A \to B}$, produced or consumed in an irreversible change of a system from state **A** to state **B** depends on the path chosen (hence neither W nor Q are state functions). Consider, for instance, a fresh battery (state **A**) which is discharged (state **B**). If the electric current passes through a motor, it is possible to recover work, and little heat is produced or absorbed from the surroundings. On the other hand, if the battery is discharged through a passive resistor, only heat is produced and no work. Although the initial and final states of the battery are the same for the two cases, $W_{A \to B}$ and $Q_{A \to B}$ depend on the path chosen.

For an isolated system, $W = 0$, $Q = 0$, by definition. Hence, $\Delta E = 0$, i. e., energy is conserved when an isolated global system contains two subsystems, 1 and 2, which interact without transfer of work. According to Eq. (I-6), the energy, E_G, of the global system remains constant:

$$\Delta E_G (= \Delta E_1 + \Delta E_2) = 0 \tag{I-7}$$

and

$$-Q_{1 \to 2} = Q_{2 \to 1}. \tag{I-8}$$

It is seen that the heat given up by subsystem 1 is equal to the heat received by 2 (or vice versa). Hence, one may define a uniform flow of heat, $\mathscr{I}_{Q'}$ from one system to the other. *Heat transfer*, measured in terms of this flow, is frequently considered in this text.

These considerations all refer to the fundamentals of *macroscopic* thermodynamics. *Molecular* thermodynamics, based on the mechanical theory of heat, assumes that heat flow amounts to energy transfer on the molecular scale. This assumption, which has proven enormously fruitful in the development of science, makes, by its very nature, caloric heat flow, Q_{cal}, equivalent to thermodynamic heat flow, Q_{th}, and hence, by definition, to energy transfer. Therefore, the whole of Eq. (I-6) became the law of energy conservation, often called the first law of thermodynamics, although it was established in its entirety after the second law of thermodynamics had been found.

Problems

I.1. a) Prove Archimedes' principle from the energy-conservation principle. Archimedes' principle states that "a body immersed in a liquid experiences an apparent loss of weight which is equal to the weight of the liquid displaced."
b) Clay [specific weight $\rho_a = 1.10$ g cm^{-3}, specific heat $\bar{c}_a = 0.85$ cal g^{-1} (K)$^{-1}$] is added to 100 cm^3 of a saline solution [$\rho_{sol} = 1.05$ g cm^{-3}, $\bar{c}_{sol} = 0.98$ cal g^{-1} (K)$^{-1}$] so as to form a 10% (by volume) suspension. The suspension is well shaken at 25 °C and then left to stand in an insulated container. As the clay settles, the average drop of the clay particles, Δh, is -25 cm.

Calculate:

1. the change, if any, of the total energy, ΔE, of the clay-water system;
2. the temperature change, ΔT, resulting from the settling process;
3. the entropy change, ΔS, due to this temperature change only.
Given: Gravitational acceleration $g = 980.7$ cm s^{-2}
       Conversion factor: 4.186 J ($= $ W s) cal^{-1}
       g stands for gram.

I.2. Prove the following thermodynamic relationships, assuming only pressure-volume work is exchanged between system and surroundings:

a) $\left(\dfrac{\partial \mu_i}{\partial T}\right)_{p,n} = -\bar{S}_i$

b) $\left(\dfrac{\partial \mu_i}{\partial p}\right)_{T,n} = \bar{V}_i$

c) $\left[\dfrac{\partial(\mu_i/T)}{\partial T}\right]_{p,n} = -\dfrac{\bar{H}_i}{T^2}$

d) $\left(\dfrac{\partial p}{\partial T}\right)_V = \left(\dfrac{\partial S}{\partial V}\right)_T$

e) $\left(\dfrac{\partial V}{\partial T}\right)_p = -\left(\dfrac{\partial S}{\partial p}\right)_T.$

Overbars indicate partial molal quantities. Subscript n indicates that the chemical composition of the system remains constant.
    Methods of classical equilibrium thermodynamics are sufficient for the proof of these relationships which are of importance in subsequent chapters.

I.3. One mol of an ideal gas expands from 1 atm, 350 K to 0.8 atm, 330 K. Calculate the changes of energy, enthalpy and entropy.
    Given the molar specific heats at constant pressure, $\tilde{C}_p = 7.00$ cal mol^{-1} (K)$^{-1}$ and at constant volume, $\tilde{C}_v = 5.00$ cal mol^{-1} (K)$^{-1}$ which are practically independent of the temperature in this temperature range.

1.4. The length, l, of an elastic fiber is related to the stretch force, f, by the equation of state (Zemansky 1968):

$$f = K'T \left(\frac{1}{l_0} - \frac{l_0^2}{l^2} \right).$$

K' is a constant and l_0 is the length at zero tension. l_0 *is a function of the (absolute) temperature, T.* If the fiber is stretched isothermally and reversibly from $l = l_0$ to $l = 2l_0$,
a) calculate the work necessary;
b)[15] show that the heat flow to the surroundings is

$$-Q = K'T l_0 (1 - \tfrac{5}{2} \alpha_0 T),$$

where α_0 is the (linear) thermal expansivity at zero tension:

$$\alpha_0 \equiv \frac{1}{l_0} \frac{dl_0}{dT},$$

α_0 is taken independent of the temperature.
c) What is the (total) energy gain of the fiber? Neglect the change of fiber volume with fiber length.
d) Note that, unlike many metal fibers, the total energy gained c) by this elastomer does not equal the work invested in the stretching process a). How does the energy change of an ideal gas upon isothermal compression compare to the compression work?

1.5. a) Calculate the molar enthalpy change, $\Delta \bar{H}$, the entropy change, $\Delta \bar{S}$, and the Gibbs free energy change, $\Delta \bar{G}$, resulting from the freezing at $-10°C$ of 1 mol of supercooled water at 1 atm. For water and ice, the specific heats, $\bar{c}_p$, are almost temperature-independent, and equal to 1.00 and 0.48 cal g^{-1} (K)$^{-1}$, respectively. The heat of fusion of ice at 0°C is 80 cal g^{-1}.
b) If the water is frozen while in contact with a large reservoir at $-10°C$, calculate ΔH, ΔS, and for the reservoir. (Assume that the pressure remains at 1 atm, and volume changes in the reservoir are negligible.)

---

15 Hint: To solve (I.4.b) use the equation

$$\left(\frac{\partial f}{\partial T} \right)_l = - \left(\frac{\partial S}{\partial l} \right)_T,$$

which is analogous to Eq. (d) in Problem (I.2). The minus sign is due to the fact that work must be done to *compress* a system of a given volume V, whereas the *stretching* of the fiber considered here requires work.

c) Calculate ΔH, ΔS, and ΔG as well as the exergy change, $\Delta \Lambda$, for the *global* system (water + reservoir).

Hint: Note that in order to calculate $\Delta \bar{S}$ and $\Delta \bar{G} \equiv \Delta \bar{H} - T\Delta \bar{S}$, it is necessary to devise a sequence of *reversible* processes to transform the system from its initial to its final state, e. g., water ($-10°C$) → water ($0°C$) → ice ($0°C$) → ice ($-10°C$).

Selected Literature

Textbooks of (Primarily) Equilibrium Thermodynamics

Baehr HD (1966) Thermodynamik. 2nd edn, Springer, Berlin-Heidelberg-New York

Jones JB, Hawkins GA (1963) Engineering thermodynamics. John Wiley and Sons, New York

Lewis GN, Randall M (1961) Thermodynamics. 1923 revised by Pitzer KS, Brewer L, 2nd edn, McGraw-Hill, New York

Denbigh K (1971) The principles of chemical equilibrium. 3rd edn, Cambridge University Press

Keenan JH (1941) Thermodynamics. John Wiley and Sons, New York

Hatsopoulos GN, Keenan JH (1965) Principles of general thermodynamics. John Wiley and Sons, New York
This text is in two parts, viz., an introduction for the beginner, followed by an advanced general approach, based on the writings of Gibbs. In their entirety, the two parts form an extraordinary reference work.

Noyes AA, Sherill MS (1938) A course of study in chemical principles. 2nd edn, Macmillan, New York
This text has a multitude of problems, which are an integral part of the presentation. The concept of *reversibility* is explained in considerable detail.

Obert EF (1960) Concepts of thermodynamics. McGraw-Hill, New York

Prigogine I, Defay R (1950) Thermodynamique chimique. 1st edn, Desoer, Liège

Bruhat G (1968) Thermodynamique. 6th edn, Masson, Paris

Souchay P (1964) Chimie générale (thermodynamique chimique). Masson, Paris

Fabry A (1930) Éléments de thermodynamique. Librairie A Colin, Paris

Campbell JA (1970) Chemical systems. WH Freeman and Co, San Francisco
Appendix III of this text is a compilation of 22 statements of the second law of thermodynamics.

Dickerson RJ (1969) Molecular thermodynamics. WA Benjamin Co, Palo Alto California

Guggenheim EA (1967) Thermodynamics. 5th edn, North Holland Publishing Co, Amsterdam

Zemansky MW (1968) Heat and thermodynamics. An Intermediate Textbook, 5th edn, McGraw-Hill, New York

Wukalowitsch MP, Nowikow II (1962) Technische Thermodynamik. German version of the second Soviet edition, edited by Elsner N, Fratzscher W, Köhler K, VEB Fachbuchverlag, Leipzig
This text has an extensive section (by Fratzscher W) on the history and importance of the exergy concept in engineering thermodynamics.

Redlich O (1976) Thermodynamics. Elsevier, Amsterdam
This is an advanced text with emphasis on unambiguous definitions of fundamental concepts.

Textbooks of Non-Equilibrium Thermodynamics

Wiśniewski J, Staniczewski B, Szymanic P (1976) Thermodynamics of nonequilibrium processes. D Reidel Publishing Co, Dordrecht (Holland)
Prigogine I (1967) Introduction to thermodynamics of irreversible processes. 3rd edn, Wiley-Interscience, New York
Denbigh KG (1951) The thermodynamics of the steady state. Methuen & Co, London
Van Rysselberghe P (1963) Thermodynamics of irreversible processes. Hermann, Paris
De Groot SR (1961) Thermodynamics of irreversible processes. North-Holland Publishing Co, Amsterdam 4th printing
Katchalsky A, Curran P (1965) Nonequilibrium thermodynamics in biophysics. Harvard University Press, Boston
Miller DG (1960) Thermodynamics of irreversible processes. Chem Rev 60:15
Haase R (1969) Thermodynamics of irreversible processes. Addison Wesley, Reading Mass
Tykodi RJ (1967) Thermodynamics of steady states. Macmillan, New York
Keller JU (1977) Thermodynamik irreversibler Prozesse. de Gruyter, Berlin
Chartier P, Gross M, Spiegler KS (1975) Applications de la thermodynamique du non-équilibre; bases d'énergétique pratique. Hermann, Paris
Glansdorff P, Prigogine I (1971) Thermodynamic theory of structure, stability and fluctuations. Wiley Interscience, New York
    This is an advanced treatise dealing with *non-linear* irreversible processes.

Referenes on Milestones in the Development of Energetics

Maxwell JC (1878) Matter and Motion. D Van Nostrand Co, New York
Maxwell JC (1888) Theory of heat. 9th edn, Longman's, Green & Co, London
    Available energy (exergy) is discussed on pp 185 – 193
Gibbs JW (1973) Trans Conn Acad Arts Sci II:382
    (Re-edited by Yale University Press, New Haven Conn 1948 "The Collected Works of J Willard Gibbs," Vol I)
    Definitions of reservoir ("medium"), energy (meaning *total* energy) and *available* energy are found on pp 40, 50 and 53, respectively.
Keenan JH, Shapiro AH (1947) History and exposition of the laws of thermodynamics. Mechan Engin 69:915
Loebl EM (1960) J Chem Ed 37:361

History of the Exergy Concept

Haywood RW (1974) A critical review of the theories of thermodynamic availability with concise formulations. J Mechan Engin Sci (I Mech E) 16:160
    This article is in two parts, the second of which concludes with a critical historical review of exergy.
Fratzscher W (in German) Appendix to the text by Wukalowitsch and Nowikow cited above in the selected list of textbooks on equilibrium thermodynamics
Evans RB, Crellin GL, Tribus M (1980) Ch I In: Spiegler KS, Laird ADK (eds) Principles of desalination. 2nd edn, Academic Press, New York
Rant R (1956) Forsch Geb Ingenieurwes 22(1):36
Rant R (1957) Allg Wärmetech 8(2):25
Van Lerberghe G, Glansdorff P (1932) Le rendement maximum des machines thermiques. Extraits des publications de l'Ass des Ingénieurs de l'École des Mines de Mons 42
Gouy M (1889) Sur l'énergie utilisable. J Phys Théor Appl Sér 2ᵉ 8:501
Evans RB (1969) A proof that essergy is the only consistent measure of potential work. Ph D Thesis, Thayer School of Engineering, Dartmouth College, Hanover, New Hampshire

Chapter II. Exergy

A. Introduction

The purpose of this chapter is (a) to express the available energy (exergy) of systems (i.e., the maximum work obtainable, given a specified environment) in terms of familiar parameters, which can often be found in thermodynamic tables, and (b) to relate the loss of exergy in irreversible processes to the concomitant increase of entropy. Also, since the loss of exergy from a given system usually translates into a loss of monetary value of the system, elementary considerations about the relations between exergy and capital costs of processes (thermoeconomics) are introduced.

B. Exergy of a Closed System

1. Isothermal Processes

Classical thermodynamics demonstrates that when a system changes isothermally and at a constant pressure from one state to another, the maximal useful work, $W_{u,rev}$, which can be gained from this transformation is the difference between the initial and final values, respectively, of the total *Gibbs free energy*, $G \equiv H - TS$,[1] of the system. Similarly, for a change at constant volume, the maximal work is the difference between the system's initial and final Helmholtz function (also called Helmholtz free energy), $F_{Helm} \equiv U - TS$. Both free energies are *state functions*, because the parameters U, H, T and S depend only on the state of the system rather than on the specific path chosen to bring it to the given state (as shown in Sect. I.C, Postulate 3, W and Q are *not* state functions).

Consider, for instance, the discharge of a battery at constant temperature and pressure, as shown in Fig. I-1 (the battery is in contact with a reservoir, i.e., a thermostat). This production of electric power is possible because of the electrode reactions, the sum of which represents the chemical reaction which gives

---

1 As shown in Sect. II.B.3, G is identical with the exergy Λ, when system and reservoir temperatures and pressures are the same.

rise to this power production. The initial (charged) state of the battery is designated by subscript 1, and the final (discharged) state by subscript 2. The volume change of the battery is ΔV.

If the discharge is carried out under reversible conditions, i.e., at very low current, the voltage of the battery, and hence the electric power produced, are higher than in a rapid, irreversible discharge, as discussed in Sect. I.B. The maximum *useful* work. $W_{u,rev}$ (i.e., the total work minus the pressure-volume work, $p\Delta V$, done by the battery against the reservoir pressure, as a result of the expansion of the battery) equals $G_1 - G_2 = -\Delta G$. This maximal useful work is electrical. In general, a certain amount of heat, Q, must flow from or to a reservoir at temperature T, even when the discharge proceeds reversibly, if the temperature of the battery is indeed to be held constant at T. It should be noted that this heat flow is not equal to the *heat of reaction* which is the heat evolved when the chemical reaction proceeds without production of electric current. The heat of reaction which equals the difference between the enthalpies of the reactants and reaction products, respectively (i.e., $-\Delta H_u$), can be measured by mixing the components in a vessel contained in a calorimeter, allowing the chemical reaction (e.g., $H_2 + Cl_2 \rightarrow 2HCl$, see Fig. I-1) to proceed without production of electric power and hence not necessarily reversibly.

During the isothermal and isobaric reversible discharge, the total enthalpy change of the battery is $\Delta H_u \equiv \Delta U + p\Delta V = Q - W_{u,max}$ [Eq. (I-6)]. Since this maximum useful work equals $-\Delta G$, the total enthalpy change of the battery is $Q + \Delta G$. Frequently, some heat flows from the battery to the reservoir during this discharge (Q < 0). In that case

$$(\Delta H_u) < (G_2 - G_1) \equiv \Delta G \quad (Q < 0, \Delta H_u < 0, \Delta G < 0) . \tag{II-1}$$

Occasionally, however, the battery absorbs heat from the reservoir during the discharge, especially when the reaction proceeds at high temperature. For instance, at temperatures above 750°C the reaction $2C_8H_{18} + 25O_2 \rightarrow 16CO_2 + 18H_2O$ absorbs heat when performed reversibly (Q > 0), which means that the enthalpy change, ΔH_u, is less negative than the Gibbs free-energy change, ΔG, at these high temperatures (Dugdale 1965, p 27).

2. Non-Isothermal Processes; the Exergy Concept for Static Systems[2]

The question now arises if the maximum useful work, $W_{u,rev}$, which can be gained from a *non-isothermal* change can be readily determined from tabulated

---

2 In addition to the references on exergy listed in Chap. I, readers are referred to the article by Glansdorff 1957, and to the brochure *Energie und Exergie*, Verein deutscher Ingenieure (VDI) Verlag, Düsseldorf, 1965.

data, similar to the calculation for an *isothermal* change by the use of the function G listed in thermodynamic tables for many substances under a variety of conditions. (Conversely, if the change of the system requires the investment of work, $W_{u,rev}$ is negative and represents the *minimum* work which must be expended to produce the desired change.) In particular, it is assumed that the thermostat or reservoir (which is at our disposal for heat rejection or uptake from and to the system, respectively) is not necessarily at the same temperature as the system. It will be shown in the following that even in this case, $W_{u,rev}$ can indeed be quantitatively expressed in terms of customary thermodynamic state functions of the system, plus the temperature, T_0, and the pressure, p_0, prevailing in the reservoir. This type of *monothermal* process is of importance for processes taking place in a uniform environment, e.g., in quiet air, or in a uniform and quiet body of water.

a) Maximum Work, W_{max}

To estimate the *maximum* work, dW_{max}, for an infinitesimal monothermal change of the system A (Fig. II-1), in which the heat dQ_0 flows from A (temperature T_A), and the reservoir (at temperature T_0) receives the heat dQ_0, it is necessary to think of a *reversible* process for the withdrawal of heat from system A. In practice, heat-flow processes are not reversible. For instance, if A is a high-temperature system at 800°C and heat were to flow from it across a finite temperature difference of 775°C to a reservoir at 25°C, the flow would not be reversible, because small temperature changes of either system or reservoir would not reverse the direction of the heat flow. Heat removal from A under reversible conditions is possible, however, if we imagine a small reversible engine, e.g.,

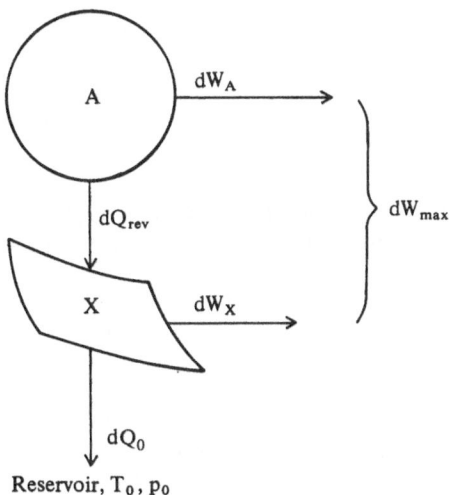

Fig. II-1. Schematic representation of reversible, monothermal heat removal from system A. X is a reversible heat engine, e.g., a Carnot engine

a Carnot engine as operating between the compound system consisting of A and the reservoir, withdrawing dQ_{rev}, from A — and, in the process, gradually cooling it — depositing a lesser quantity of heat, dQ_0, in the reservoir, and producing the work, dW_X, in its complete cycle. In this case, the process is reversible because the sequence of operations of the Carnot engine could be reversed by an infinitesimal change, i.e., instead of producing dW_X, we could introduce work into the Carnot engine, thus converting it into a heat pump. In this case, heat would flow from the reservoir to the reversible engine, and (a larger amount of) heat would be deposited in the system A.

In addition to the work produced by the Carnot engine, the system produces, or receives, work dW_A, independent of this engine. For instance, if A were a high-temperature fuel cell, it could produce a maximum amount of electrical work, dW_A, while the Carnot engine produces dW_X.

For the composite system A + X, the work is given by Eq. (I-6):

$$dW_{max} = dW_A + dW_X = -dE_A + dQ_0 . \tag{II-2}$$

The energy change of the Carnot engine, dE_X, is taken as zero, because one complete cycle of this small engine is considered here, at the end of which the engine has returned to its original state. dQ_0 is the heat received by the reservoir. When dQ_0 is positive, heat flows *out* of the composite system A + X, hence the term dQ_0 in the energy balance, Eq. (II-2), for this composite system is negative.

According to Carnot's reasoning about reversible engines, the heat flows are related by [see Eq. (A-I-15)]

$$\frac{dQ_0}{T_0} = \frac{dQ_{rev}}{T_A} . \tag{II-3}$$

Moreover, since the entropy change, dS_A, of system A is defined as

$$\frac{dQ_{rev}}{T_A} \equiv dS_A , \tag{II-4}$$

the maximum work produced by the whole system while the reversible engine X completes one full cycle can be expressed from Eq. (II-2) in terms of state functions of the system A:

$$dW_{max} = -dE_A + T_0 dS_A = -d(E - T_0 S)_A . \tag{II-5}$$

Since state functions of various systems can often be found in thermodynamic tables, or readily computed from tabulated data, Eq. (II-5) is more useful than Eq. (II-2).

b) Maximum **Useful** Work, $W_{u,max}$

The maximum *useful* work of the transformation of the system from state A to equilibrium with the reservoir is obtained by subtracting from dW_{max} the expansion work against the reservoir pressure p_0, viz., $p_0 dV_A$. (If the system contracts, dV_A is negative and the term $-p_0 dV_A$ becomes positive, i.e., the work done by the reservoir pressure is added to the total work.[3]):

$$dW_{u,max} \equiv dW_{max} - p_0 dV_A = -d(E - T_0 S + p_0 V)_A \equiv -d\Lambda_A , \qquad \text{(II-6)}$$

which indicates that the *exergy* of the system A is represented by the following expression involving only state functions:

$$\Lambda \equiv E - T_0 S + p_0 V + \text{const} . \qquad \text{(II-7)}$$

The useful work obtained when the system changes from state **1** to state **2** by a reversible path is the *maximum* useful work obtainable for a transition from the given initial state **1** to the given final state **2**.

$$W_{u,max} = \int_1^2 dW_{u,max}$$

$$= -\int_1^2 d\Lambda = -\,{}_1^2(\delta\Lambda) = \Lambda_1 - \Lambda_2 . \qquad \text{(II-8)}[4]$$

This integral depends only on the nature of the initial and final states, **1** and **2**, respectively, rather than on the integration path, because the parameters E, S and V in the expression for the exergy [Eq. (II-7)] are state functions and hence independent of the path by which the system reaches its given state; the remaining two parameters, T_0 and p_0, are reservoir constants which depend only on the choice of the reservoir and again not on the path for the change of the system from state **1** to state **2**.

The independence of the exergy difference, $\delta\Lambda$, on the path from state **1** to **2** can also be inferred from purely thermodynamic reasoning. Consider a monothermal cycle **1** → **2** → **1**, the second step of which, **2** → **1** is reversible. If **1** → **2**

---

3 The distinction between maximum work, W_{max}, and maximum *useful* work, $W_{u,max}$, is analogous to the distinction between *absolute* and *gauge* pressure, which is common in engineering science.

4 It should be remembered that the symbols Δ and d indicate the difference between a parameter written on the right and a parameter written on the left. For instance, for the sequence $X_1 \rightarrow X_2$, $\Delta X \equiv X_2 - X_1$. δ indicates the difference between parameters in the final and initial states, respectively. For chemical reactions, Δ is also used, because it is customary to write chemical equations such that the reaction products (final state) are on the right, and the reactants (initial state) on the left.

produced more useful work than $2 \to 1$ consumes, the system would return to its original state, having produced a *net* amount of useful work by heat-flow interaction with a single reservoir. But, by the definition of a reservoir, this is impossible. Hence the step $1 \to 2$ cannot produce more useful work than $2 \to 1$ consumes. Moreover, since the second step is reversible, its consumption of useful work is numerically (almost) equal to the useful work which it would produce when reversed, and cannot be numerically less than the production of useful work in the first step. It follows that, given the reservoir properties, no monothermal process (irreversible or reversible) transforming the system from state 1 to 2 can produce more work than a *reversible* process, and that *all* reversible processes produce the same maximum amount of useful work, $W_{u,\text{max}}$. Equation (II-8) states that this maximum useful work is the exergy difference $\Lambda_1 - \Lambda_2$.

c) Complete Expression for Exergy

The constant in Eq. (II-7) remains to be found. By definition, Λ_1, is the maximum useful work obtainable when the system changes from state 1 to *equilibrium* with the reservoir. When the system reaches equilibrium with the reservoir, it is said to be completely relaxed with respect to the reservoir. It cannot produce work with or without interaction with the reservoir; therefore its pressure and temperature are p_0 and T_0, respectively. Examples of such systems in equilibrium with the reservoir are a spring which, originally hot and stretched, has been cooled to the reservoir temperature, and its tension relaxed, or a completely discharged battery at the temperature and pressure of the ambient air.

In this final state, $E = E_0$, $S = S_0$ and $V = V_0$. Considering this state as state 2 in Eq. (II-8), we obtain from the constant in Eq. (II-7)

$$\text{const} = -(E_0 - T_0 S_0 + p_0 V_0)_A \tag{II-9}$$

and hence, dropping, for convenience, the subscript A from Λ, E, S and V:

$$\Lambda = (E - E_0) - T_0(S - S_0) + p_0(V - V_0). \tag{II-10}$$

3. Some Properties of Exergy

a) E is the *total* energy of the system, as defined in Chap. I. It includes kinetic and potential energy. Hence the exergy [Eq. (II-10)] also contains all these terms. A system can produce or consume work by change of its velocity (kinetic work), position in a field, e.g., an electric or gravity field (work due to a change of potential), linear or surface tension, electrochemical reactions, etc.

In a simple system at rest, ΔE may be replaced by the change of *internal* energy, ΔU.

b) According to Eq. (II-8) the maximum useful work obtainable from a system changing monothermally from state **1** to state **2** is the exergy difference $-\Delta\Lambda$:

$$W_{u,\,max} = -{}^2_1(\Delta\Lambda) = \Lambda_1 - \Lambda_2. \qquad\qquad\text{(II-11)}$$

This is true whether the change is isothermal and/or isobaric, or not. For the special case of an isothermal and isobaric change, in which the reservoir (thermostat) temperature, T_0, and pressure, p_0, are identical with temperature, T, and pressure, p, prevailing in the system, respectively, $\Delta\Lambda$ is identical with the change of the Gibbs free energy, ΔG. It is seen that Λ is a more general function than G. Note that the numerical values of both Λ and G depend on the reference (reservoir) state chosen. For non-isothermal, non-isobaric transformations, $\Delta\Lambda$ takes the place of ΔG, which is frequently of great importance for isothermal, isobaric reactions.

c) Both entropy and exergy are preserved in global systems only if *all* processes in the global system are reversible. The inequality (I-4) for the entropy change in a closed system:

$$d_i S > 0 \qquad\qquad\text{(I-4)}$$

can also be expressed in terms of the exergy change

$$-d\Lambda > dW_u \quad \text{or} \quad -d\Lambda = dW_u - \left(\frac{\partial_i\Lambda}{\partial t}\right) dt = dW_u - d_i\Lambda. \qquad\text{(II-12)}$$

$(d_i\Lambda/dt) < 0$ is the rate of exergy change in a system undergoing an irreversible change. Both equations (I-4) and (II-12) refer to a global, isolated system. It is quite possible that the entropy of a subsystem decreases even during an irreversible transformation of the global system; but an entropy increase in another part of the system more than compensates for this entropy decrease. Similarly, the exergy may increase in a specific location, but the exergy destruction elsewhere in the global system is numerically larger.

It will be shown that the destruction of exergy in irreversible processes is related to the concomitant creation of entropy by a simple relation (theorem of Maxwell-Gouy-Stodola, Sect. II.C).

C. Exergy Accounting [5] in Flow Systems

1. Flow Systems

Consider a control volume within well-defined boundaries, e.g., a turbine or even a complex system such as a complete factory. Mass, heat and work flows may occur across these boundaries. Such a control volume, schematically shown in Fig. II-2, is called a flow system. Much of the engineering literature also uses the term open system.

The analyses in the remainder of this chapter are for the *steady states*, i.e., states in which the local temperatures, pressures, concentrations of each species, partial molal energies, and all other *intensive parameters at any spot do not change with time* in spite of the transport of mass, heat and/or work across boundaries of the flow system. The extensive properties of the flow system, e.g., mass, total energy, entropy, exergy and volume, remain always constant as long as the steady state prevails. Inflow and outflow of mass at the system boundaries are neither accelerated nor decelerated.

All *incoming* flows, $\mathcal{J}'$ (shown on the left in Fig. II-2), are counted positive, whereas all exiting flows, $\mathcal{J}''$, are counted negative, except for work flows, $\mathcal{J}_{W'}$, for which the opposite sign convention is used, in accordance with tradition and with the sign convention followed in Chap. I for closed systems. [Fig. II-2 shows only a single (exiting) work flow.[6]]

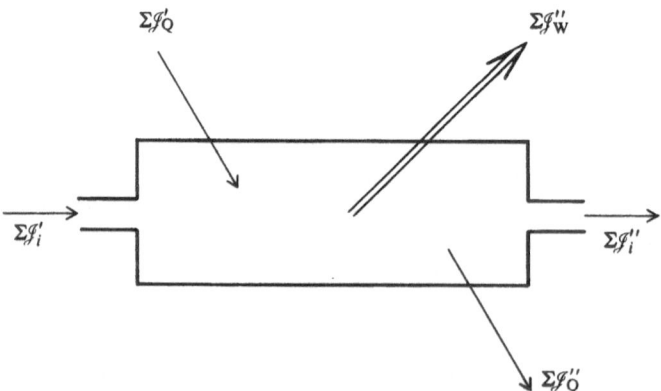

Fig. II-2. Schematic representation of a flow system

---

5 The term accounting, rather than balance is used in this book when the parameter dealt with is not conserved, e.g., exergy (or entropy) in irreversible processes.

6 Note that $\mathcal{J}$ represents a *flow*, i.e., the quantity of heat, work or of a given (mass) component entering the system per second. On the other hand, J designates a flux, i.e., a flow per unit area. If the surface is $\mathfrak{a}$, and if the flux is uniform, then $\mathcal{J} = J\mathfrak{a}$.

When the mass flows, $\mathscr{I}_i$, do not react chemically, e.g., if the system is a steam turbine, or simply a valve in a fluid line, then, in the steady state: $\mathscr{I}_i' = -\mathscr{I}_i''$. If a chemical reaction takes place, on the other hand, the values of the entering reactant flows, $\mathscr{I}_A'$, etc. are evidently not necessarily numerically equal to the flows of the exiting reaction products, $\mathscr{I}_C'$, etc. For reactions proceeding to completion, $\mathscr{I}_A'' = 0$. The temperature of the region of the system in which a heat flow enters is designated as T', while the temperature of the system in the region of an exiting heat flow is T''.

2. Energy Balance

Since the total energy of the control volume remains constant in the steady state, it is of interest to begin this analysis with an energy balance, stating that the algebraic sum of all energy flows is zero. If the heat flows are expressed in the same units as the work flows, e.g., in watts, no unit conversion is necessary, when entering both flows into this sum. As for the mass flows, it follows from Postulate 2 that each mass carries a definite amount of total energy. Thus, the total rest energy assigned to each mass flow, $\mathscr{I}_i$ (mol s^{-1}), is $\mathscr{I}_i\bar{E}_i$, where $\bar{E}_i$ is the partial molar energy, W s mol^{-1} (1 W s = 1 J). Any incoming mass flow also carries, in addition to its rest energy, the energy corresponding to the work expended to push it into the control volume, namely $\mathscr{I}_i' p' \bar{V}_i'$ W s mol^{-1}. Here, $\bar{V}_i'$ is the partial molar volume of component i at the point of entry into the control volume (the pressures on the two sides of the entrance into the control volume are virtually equal, since the flow into the system does not undergo acceleration or deceleration). Hence, the total energy introduced per second by the flow of mass i is

$$\mathscr{I}_i'(\bar{E}_i' + p'\bar{V}_i') = \mathscr{I}_i'\bar{H}_i' . \tag{II-13}$$

$\bar{H}_i'$ is the total partial molar enthalpy of i in the entrance state (').

Conversely, to the rest-energy flow, $\mathscr{I}_i''\bar{E}_i$, of each exiting stream, we add the work, $\mathscr{I}_i'' p'' \bar{V}_i''$, which this stream has to carry in order to displace the fluid in the exit line so as to maintain the steady state. Hence, each exiting stream carries $\mathscr{I}_i''(\bar{E}_i'' + p''\bar{V}_i'') = \mathscr{I}_i''\bar{H}_i''$ energy units per second.

Therefore, the energy balance (per second) of the system is

$$\sum \mathscr{I}_E' + \sum \mathscr{I}_E'' = \sum \mathscr{I}_i'\bar{H}_i' + \sum \mathscr{I}_i''\bar{H}_i'' + \sum \mathscr{I}_Q' + \sum \mathscr{I}_Q'' - \mathscr{I}_W'' = 0 . \tag{II-14}$$

Had there been incoming *and* outgoing work flows, the simple term $-\sum \mathscr{I}_W''$ would be replaced by $-(\sum \mathscr{I}_W' + \sum \mathscr{I}_W'')$.

3. Enthalpy Balance in Adiabatic Flow Systems

Consider a well-insulated turbine ($\mathscr{I}'_Q = 0 = \mathscr{I}''_Q$) driven by a single working fluid, i, and without chemical reaction. In this case, $\mathscr{I}'_i = -\mathscr{I}''_i \equiv \mathscr{I}_i$. Eq. (II-14) shows that in this case the work produced per mol of working fluid passing through the turbine is simply the difference between the molar enthalpies of incoming and outgoing flows, respectively, of working fluid:

$$\left(\frac{\mathscr{I}''_W}{\mathscr{I}_i}\right)_{\Sigma\,\mathscr{I}_Q = 0\,=\,\Sigma\,\mathscr{I}'_Q} = \bar{H}'_i - \bar{H}''_i . \qquad (\text{II-15})$$

For a well-insulated valve ($\mathscr{I}'_Q = 0$, $\mathscr{I}''_Q = 0$, $\mathscr{I}''_W = 0$), Eq. (II-14) reduces to a very simple form, viz.,

$$\bar{H}'_i = \bar{H}''_i \quad \text{for} \quad \Sigma\,\mathscr{I}_Q = 0 , \quad \Sigma\,\mathscr{I}_W = 0 . \qquad (\text{II-16})$$

It is seen that enthalpy is conserved in this case; this is true whether the processes taking place in the turbine be reversible or not.

This type of *enthalpy balance* is very useful for the analysis of various engines or processes, and is usually the very first step in such energetic analyses practiced in courses of engineering thermodynamics.

4. Exergy Accounting

Equation (II-10), which expresses the exergy of a system in terms of tabulated properties of the system (e. g., E and S) in its given state, and after it has reached equilibrium with the reservoir (e. g., E_0 and S_0), contains the term $-T_0 S$. Since, in global systems, entropy is conserved only in reversible processes, while being created in irreversible processes (Clausius' theorem), it is seen that for constant-volume systems, exergy is conserved only in reversible processes also. It follows that only when all processes taking place in a control volume (Fig. II-2) in the steady state are reversible, is the sum of the exiting exergies numerically equal to the sum of the incoming exergies. In the absence of heat flows this means that in such a system the following relation holds:

$$\Sigma\,\mathscr{I}'_{A_i,\text{rev}} + \Sigma\,\mathscr{I}''_{A_i,\text{rev}} - \Sigma\,\mathscr{I}_W = 0 . \qquad (\text{II-17})$$

Here, the flows, $\mathscr{I}'_{A_i,\text{rev}}$, represent the exergies entering the control volume (which are counted positive) and the $\mathscr{I}''_{A_i,\text{rev}}$ represent the exergy flows leaving the control volume, counted negative. The sum of the work flows, $\Sigma\,\mathscr{I}_W$, is a pure exergy flow, and therefore enters the exergy accounting directly in Eq. (II-17). Outgoing

and incoming work are counted positive and negative, respectively. Hence, the
negative sign before $\sum \mathcal{J}_W$.

The flows of the exergies, $\mathcal{J}_{A_i}'$, brought in with the mass flows, $\mathcal{J}_i'$, are calculated by a reasoning similar to that used in the calculation of the enthalpy
flows (Sect. II.C.3). In other words, we add to the rest exergy the *useful* work
necessary to push the mass flow into the control volume. Since this work could
conceivably be recovered from mass flow by means of a turbine, the useful work
so invested represents indeed a contribution to the exergy of the incoming mass.
For each mol of inflowing mass i, this additional exergy is equal to $(p' - p_0)\bar{V}_i'$.
Adding this term to the expression for the rest exergy [Eq. (II-10)], we obtain

$$\mathcal{J}_{A_i}' = \mathcal{J}_i' [(\bar{E}_i' - \bar{E}_{i,0}) + p_0(\bar{V}_i' - \bar{V}_{i,0}) - T_0(\bar{S}_i' - \bar{S}_{i,0}) + (p' - p_0)\bar{V}_i'] . \quad \text{(II-18)}$$

The overbars denote partial molar properties of component i.

Since, by definition, the partial molar enthalpy of i is

$$\bar{H}_i' \equiv \bar{E}_i' + p'\bar{V}_i' \quad \text{(II-19)}$$

it follows that Eq. (II-18) can be presented in the simpler form

$$\mathcal{J}_{A_i}' = \mathcal{J}_i' [(\bar{H}_i' - T_0\bar{S}_i') - (\bar{H}_{i,0} - T_0\bar{S}_{i,0})] . \quad \text{(II-20)}$$

Hence, the *flow exergy*, $\bar{\Lambda}_i'$, brought into the control volume with each mol of
component i is

$$\bar{\Lambda}_i' = \frac{\mathcal{J}_{A_i}'}{\mathcal{J}_i'} = (\bar{H}_i' - T_0\bar{S}_i') - (\bar{H}_{i,0} - T_0\bar{S}_{i,0}) . \quad \text{(II-21)}$$

The expressions for the *exiting* exergy flows, $\mathcal{J}_{A_i}''$, and for the partial molar
flow exergies, $\bar{\Lambda}_i''$, carried by these flows are analogous to Eqs. (II-20) and
(II-21).[7]

For isothermal transformations in which initial and final temperature of the
system are equal to the reservoir temperature $(T' = T_0 = T'')$, it is seen from
Eq. (II-21) and the definition of G (Sect. II.B.1) that the exergy difference is
equal to the Gibbs free-energy difference:

$$\sum \bar{\Lambda}_i'' - \sum \bar{\Lambda}_i' = \sum \bar{G}_i'' - \sum \bar{G}_i' . \quad \text{(II-22)}$$

---

7 The exergy nomenclature found in the literature is not uniform. For instance, the term
"availability" is sometimes used to designate the exergy expression derived for *closed*
systems [Eq. (II-10)], as well as the expression for *flow systems* [Eq. (II-21)]. Moreover,
it is not always clear whether the literature expressions include *all* forms of useful work,
including the work which could be furnished by chemical reactions performed in
suitably constructed batteries.

Energy and/or enthalpy balances for flow systems make it possible to calculate the work flow obtainable from (or invested in) the system per unit mass passing through the system. As seen from Eq. (II-15), only the terminal conditions of the working fluid (i. e., its conditions at the entrance and the exit points) have to be known to determine this work flow. On the other hand, exergy accounting gives important additional information on the exergetic efficiency of the system, i. e., the degree or reversibility of the process taking place in it. Only when all processes are reversible is exergy conserved and hence the work flow from the system equal to the difference between incoming and outgoing exergy flows. Hence exergy expresses the *quality* of the energy supply.

Since exergy carries a price tag, exergy accounting is thus useful for the *economic* evaluation of processes or equipment components. Moreover, exergy accounting also determines the efficiency of biological mechanisms. It has been postulated that the degree of this efficiency is related to species survival (for early references, see Lotka 1922).

The following section demonstrates that the rate of *exergy destruction* in an irreversible process is equal to the rate of *entropy creation* multiplied by the reservoir temperature.

D. Relationship Between Exergy Disappearance and Entropy Creation

1. Exergy Disappearance in Isolated Systems

Consider processes − not necessarily reversible ones − taking place in an isolated system of fixed volume. Its exergy change $\Delta \Lambda \equiv$ exergy of the system after the processes took place − exergy of the system before the processes started, is, from Eq. (II-10):

$$\Delta \Lambda = \Delta E + p_0 \Delta V - T_0 \Delta S . \tag{II-23}$$

Since the system is isolated, its total energy is conserved, i. e., $\Delta E = 0$, and since the volume is fixed, $\Delta V = 0$. Hence

$$\Delta \Lambda = -T_0 \Delta S . \tag{II-24}$$

Dividing by the time of change, Δt, we obtain

$$\Delta \Lambda / \Delta t = -T_0 (\Delta S / \Delta t) \tag{II-25}$$

or, in differential form,

$$d\Lambda/dt \equiv \dot\Lambda = -T_0(dS/dt) = -T_0\dot{S}. \tag{II-26}$$

2. Exergy Changes Associated with Heat Flows, $\dot\Lambda_Q$

The exergy associated with a given heat flow is found by calculating the maximum amount of work which this heat flow could produce. This conversion can be done by imagining that a reversible engine, e. g., a Carnot engine, receives the heat flow at the temperature, T', at which it enters the system, rejects some heat into the reservoir (temperature T_0), and converts the remainder to work. According to Appendix I, the efficiency of the conversion is $1 - (T_0/T')$. If the heat flow is multiplied by this efficiency, the maximum amount of work which can be produced per second is obtained:

$$\Sigma\, \mathcal{J}'_{\Lambda_Q} = \Sigma\, \mathcal{J}_Q \left(1 - \frac{T_0}{T'}\right). \tag{II-27}$$

The exergy flow associated with the *outgoing* heat flow is calculated in similar manner:

$$\Sigma\, \mathcal{J}''_{\Lambda_Q} = \Sigma\, \mathcal{J}''_Q \left(1 - \frac{T_0}{T''}\right). \tag{II-28}$$

The rate of exergy change occurring in the system as a result of the heat flows is:

$$-\dot\Lambda_Q = \Sigma\, \mathcal{J}'_Q \left(1 - \frac{T_0}{T'}\right) + \Sigma\, \mathcal{J}''_Q \left(1 - \frac{T_0}{T''}\right). \tag{II-29}$$

3. Exergy Disappearance in Flow Systems

Consider the flow system shown in Fig. II-2, in which irreversible processes can take place. Mass flows, $\mathcal{J}_i$, heat flows, $\mathcal{J}_Q$, and work flows, $\mathcal{J}_W$, enter and leave the system.

In the stationary state, the exergy contained within the control volume is constant. The rate of exergy outflow is not necessarily equal to the rate of exergy inflow, however. The rate of exergy change between incoming and outgoing flows, $(\Lambda' - \Lambda'')/\Delta t = -\Delta\Lambda/\Delta t = -\dot\Lambda$, can be conceptually broken down into three contributions, viz., changes in the exergies associated with the mass, heat and work flows, respectively, which are quantitatively expressed in the following:[8]

$$-\dot{A} = \sum \mathscr{I}_i'(\bar{H}_i' - T_0\bar{S}_i') + \sum \mathscr{I}_i''(\bar{H}_i'' - T_0\bar{S}_i'')$$

$$+ \sum \mathscr{I}_Q'\left(1 - \frac{T_0}{T'}\right) + \sum \mathscr{I}_Q''\left(1 - \frac{T_0}{T''}\right) - \sum \mathscr{I}_W' - \sum \mathscr{I}_W'' \,. \qquad \text{(II-30)}$$

The first two terms on the right side of Eq. (II-30) are the difference between incoming and outgoing exergies, respectively, carried by mass flows [Eq. (II-20)]. The next two terms are the corresponding difference for the exergies carried by heat flows [Eq. (II-28)]. The last two terms represent the difference between incoming and outgoing work flows.

We rearrange the terms in Eq. (II-30):

$$-\dot{A} = \sum \mathscr{I}_i'\bar{H}_i' + \sum \mathscr{I}_i''\bar{H}_i'' + \sum \mathscr{I}_Q' + \sum \mathscr{I}_Q'' - \sum \mathscr{I}_W'' - \sum \mathscr{I}_W$$

$$- T_0\left[(\sum \mathscr{I}_i'\bar{S}_i' + \sum \mathscr{I}_i''\bar{S}_i'') + \sum \frac{\mathscr{I}_Q'}{T'} + \sum \frac{\mathscr{I}_Q''}{T''}\right]. \qquad \text{(II-31)}$$

According to the energy balance for the control volume [Eq. (II-14)], the algebraic sum represented by the first six terms of Eq. (II-28a) is zero. The remaining terms are

$$-\dot{A} = \underbrace{-T_0(\sum \mathscr{I}_i'\bar{S}_i' + \sum \mathscr{I}_i''\bar{S}_i'')}_{T_0\dot{S}_{mass}} - \underbrace{T_0\left(\sum \frac{\mathscr{I}_Q'}{T'} + \sum \frac{\mathscr{I}_Q''}{T''}\right)}_{T_0\dot{S}_{heat}}. \qquad \text{(II-32)}$$

Equation (II-29) can be conveniently rewritten by use of the symbols $\dot{S}(\equiv d_i S/dt)$ for the rate of entropy creation in the control volume and $-\dot{A}(\equiv -d_i A/dt)$ for the rate of exergy destruction:

$$-\dot{A} = T_0(\dot{S}_{mass} + \dot{S}_{heat}) = T_0\dot{S}_{total} > 0 \,. \qquad \text{(II-33)}$$

This equation, known as the Gouy-Stodola equation or Maxwell-Gouy equation, relates the rate of exergy destruction, $-\dot{A}$, to the rate of entropy creation, $\dot{S}$. Because $\dot{S} > 0$ in irreversible processes (Clausius' theorem), it follows that $\dot{A} < 0$, i.e., in irreversible processes, exergy is destroyed. The equation is

---

8 Note that incoming *mass* and *heat* flows, $\mathscr{I}'$, are counted positive, while outgoing flows, $\mathscr{I}''$, are negative; the opposite convention holds for *work* flows.

true not only for flow processes, but also for non-flow processes [Eq. (II-26)].[9] In accordance with electrical-engineering terminology, $-\dot{A}$ is sometimes called *dissipation*.

It is of interest that modern engineering science places the rate of *exergy destruction* into the focus of its energetic analyses (Baehr 1966), while an equivalent role is played by the rate of *entropy increase* in most presentations of non-equilibrium thermodynamics (Prigogine 1967). The Maxwell-Gouy-Stodola equation forms a bridge between these disciplines. It serves as the starting point for the thermoeconomic analysis of processes and components of equipment, i.e., an analysis in which the cost of lost exergy is compared to the investment necessary to reduce this loss (Evans et al. 1980). Some simple examples of thermoeconomic analysis are presented in Sect. II.E.3.

In continuous systems, e.g., in single-phase mixtures of gases or solutions (with concentration gradients of the components), it is desirable to apply Eq. (II-33) to *virtual* volume elements dx · dy · dz, rather than *macroscopic* control volumes, such as the one indicated in Fig. II-2. In this case, the following equation is obtained

$$- \operatorname{div} J_A = T_0 \dot{s} , \qquad\qquad (II\text{-}34)$$

where $\dot{s}$ is the rate of entropy creation *per unit volume* and J_A the (vectorial) "exergy flux" in W cm^{-2} at the location of the volume element. Note that J_A is *not* uniform, i.e., it varies from place to place when $\dot{s}$ is finite.

E. Exergetic and Thermoeconomic Analyses

The following examples refer to some exceptionally simple cases. For more complex cases, the reader is referred to the literature on exergetic (second-law) analysis, in the list at the end of this chapter.

1. Definition of Exergetic Efficiency

The exergetic efficiency, Φ_A, of a process (or item of equipment) is defined as

$$\Phi_A \equiv \frac{\text{Recoverable exergy}}{\text{Introduced exergy}} \leqslant 1 . \qquad\qquad (II\text{-}35)$$

---

9 Note that in stationary systems, the exergy of a given mass is expressed by Eq. (II-10), whereas the exergy of fluids in motion is different from stationary fluids, as explained in Sect. II.C.4, and expressed by Eq. (II-20), which is used for exergy accounting in flow processes.

Recoverable exergy is the exergy leaving the process. It is recoverable in principle only; in practice, it may not be readily feasible or economical to perform this recovery.

The exergetic efficiency is unity for a reversible process and decreases with increasing irreversibility. Thus, its value indicates the degree of reversibility of the process considered.

2. Exergetic Efficiency of Two Space-Heating Methods

a) Hot-Water Heating

The room temperature, T_b is 295 K. Hot water enters the radiators at $T_e = 333$ K and leaves at 295 K. The reservoir temperature is taken as the temperature of the atmosphere *outside* the building, i.e., $T_0 = 273$ K.

It is seen that calories leave the building both with the hot water and through leaks, insulation losses etc., all at 295 K. In the steady state, the sum of these outgoing heat flows is numerically equal to the heat flow entering the building (due to the hot water entering) $\mathscr{J}_Q'$:

$$\mathscr{J}_Q' = - \sum \mathscr{J}_Q'' . \tag{II-36}$$

The exergetic efficiency is calculated from Eq. (II-35) by use of the heat-flow exergy expressions (II-27) and (II-28):

$$\Phi_{A,a} = \frac{-\sum \mathscr{I}_{AQ}''}{\mathscr{I}_{AQ}'} = - \frac{\sum \mathscr{J}_Q'' \left(1 - \dfrac{T_0}{T_b}\right)}{\mathscr{J}_Q' \left(1 - \dfrac{T_0}{T_e}\right)} = \frac{(T_b - T_0) T_e}{(T_e - T_0) T_b} = 0.41 . \tag{II-37}$$

The *maximal* heat flow, $- \sum \mathscr{J}_Q''$, which is the calories delivered per second by a body at (absolute) temperature T_e to a space at temperature T_b (the environmental temperature being T_0) *when the exergetic efficiency is unity*, has been named the *thermergy* flow (Silver 1981).[10] It is seen from Eq. (II-37) that the thermergy depends on all three temperatures T_b, T_e, and T_0, respectively.

b) Electric Heating

In this case, the electric power used for the electric resistance heaters represents pure exergy. In the steady state, the energy represented by the heat outflows

---

10 In this case, the heating is done by an ideal heat pump operating between T_0 and T_b, powered by a reversible engine (e.g., Carnot's) operating between T_e and T_0.

equals numerically the heat inflow (provided both are expressed in the same units, e. g., W):

$$\mathscr{I}'_W = J''_Q \tag{II-38}$$

and the exergetic efficiency is

$$\Phi_{\Lambda,b} = \frac{\mathscr{I}''_Q \left(1 - \dfrac{T_0}{T_b}\right)}{\mathscr{I}'_W} = \frac{T_b - T_0}{T_b} = 0.075 . \tag{II-39}$$

It is seen that while in both heating methods (a) and (b), energy transfer from the heater to the room air is complete, the exergetic efficiency of the hot-water heating system is much higher than of the electric heating system.

This example was given although in practice the exergy of the room air is not recovered. (In principle, this exergy is recoverable, but it would not be practical to build engines for this purpose alone.) One can see that electric heating would be wasteful if the cost of electric power were the only item in the total cost of heating the building. It is necessary to consider the fixed costs also, however, i. e., interest, amortization, insurance etc., as well as maintenance and repair costs. Although for a given caloric amount of heating, the exergy costs during operation of an electric heating system are higher than for a hot-water heating system, the fixed costs for the electric system are less because its initial cost is less. One finds often that (for a given result) the less one spends on the acquisition and installation of a process, the more one has to spend on the exergy necessary for operation of this process. *Thermoeconomics* is an optimization method which deals with the proper balancing of these cost terms, in an effort to obtain the minimum cost of the product.

3. Application of Elementary Thermoeconomic Analysis

While thermoeconomic analysis of complex processes requires detailed knowledge, the following greatly simplified example is presented to illustrate Kelvin's rule, which deals with the minimization of the product cost by optimal balancing of exergy costs and fixed costs.

Consider an electrodialysis plant, consisting of stacks of ion-selective membranes, for the desalination of brackish water.[11] In such a plant, brackish water is split into fresh water and concentrate (brine) as a result of the passage of electric current. Although at very high current densities these plants exhibit non-linear

---

11 For details, see K. S. Spiegler: Salt-water purification, 2nd ed., Plenum New York, 1977.

behavior (i. e., the electric current is not proportional to the applied voltage), in practice, electrodialysis is performed at moderate current densities, when both the electric current and the fresh-water production rate are proportional to the applied voltage, i. e., the apparent resistance, $\mathcal{R}$ ((Ω), of the electrodialysis stacks of the plant is independent of the current density.

Is a small plant which requires little investment, but a high current density (and hence large exergy dissipation), preferable to a large plant which requires high investment, but causes relatively little exergy dissipation?

The total cost of desalination *per unit quantity of product* can be broken down into three main categories, viz.:

A. The exergy cost per unit product (say, per cm^3 of fresh water produced)

This cost is proportional to the production rate ($cm^3\,s^{-1}$), because the rate of exergy dissipation for the whole plant is the dissipation of electric power $\dot{\Lambda} = \mathcal{R}\,I^2$ (W) where I is the direct current through the plant (A). Since the production rate is proportional to I, rather than to I^2, the exergy dissipated *per cm³* of fresh water produced is proportional to I. Thus, the exergy cost (electric-energy cost) per cm^3 of fresh water produced equals aI where a is independent of the current. The constant a is proportional to the local cost of electric power.

B. The investment cost per unit product

This cost, which represents primarily the repayment of borrowed capital and interest on the loan necessary to erect the plant, is inversely proportional to the fresh-water production rate and hence also inversely proportional to the current I, because the total daily outlay for the whole plant is independent of the production rate. Hence, if the production rate can be doubled, only one half of these costs is assignable to each unit of product. Thus, the investment cost per cm^3 of fresh water produced is b/I, where b is independent of the current. The constant b increases with increasing plant cost and increasing interest rate.

C. The cost of added chemicals per unit product, and other costs which are independent of the production rate. For instance, pre-purification of the raw water and/or aftertreatment of the fresh-water product often cause additional expenditure. This cost — again per cm^3 of fresh water produced — is independent of the production rate. Hence, the cost *per cm³ product* is represented by the constant c.[12]

The *total* cost of fresh water per cm^3, Ω_w is the sum of terms (A), (B) and (C), viz.:

---

12 This constant also contains the cost for the irreducible minimum exergy required to split the raw water into fresh water and brine. This cost (per cm^3 of fresh water produced) is independent of the production rate.

$$\Omega_w = aI + \frac{b}{I} + c. \tag{II-40}$$

When this cost is plotted against the current, I, a minimum cost is obtained, the value of which can be found by calculating the first derivative of Ω_w with respect to I and setting it zero:

$$\left(\frac{d\Omega_w}{dI}\right)_{I_{min}} = a - bI_{min}^{-2} = 0. \tag{II-41}$$

From this equation, the current to be used at the minimum of total cost is obtained, viz.:

$$I_{min} = \sqrt{\frac{b}{a}}. \tag{II-42}$$

It can be readily verified that there exists indeed a cost minimum, rather than a maximum, by differentiating Eq. (II-41) with respect to I: the second derivative, $d^2\Omega_w/dI^2$ is positive.

Substitution of Eq. (II-42) in (II-40) leads to the conclusion that the minimum product cost is achieved when the current is chosen such that the exergy costs, aI, are equal to the fixed costs, b/I (Kelvin's rule):

$$\Omega_{w\,min} = \sqrt{ab} + \sqrt{ab} + c. \tag{II-43}$$

The derivation of Kelvin's rule was possible in this simplified example because in the practical operating range of electrodialysis plants, only a small error is introduced by taking the electric current proportional to the applied voltage. In several other processes, approximate proportionality between flow and driving force prevails and similar considerations may then be used.[13] Therefore, Kelvin's rule is of much broader applicability than for the mere treatment of the electrodialysis process which was presented here only by way of a simple example.

As indicated by Eq. (II-42), these simple thermoeconomic considerations are useful for determining the most economical production rate in a given plant from the manufacturer's point of view. The social and environmental costs associated with production processes, e. g., the expenses for prevention or removal of pollution and the impairment of health and/or environment, which are not always borne by the manufacturer, were not introduced into the cost equation (II-40) and in the cost-minimization [Eq. (II-42)]. In principle, the cost of established

---

13 See, for instance, Silver's approach to the optimization of the number of stills in multiple-effect distillation plants (1962).

pollution-control methods can and should be introduced into cost equations. Other factors, such as the preservation of the environment are, by their very nature, not readily quantifiable. This indicates the limitations of an exclusively thermoeconomical approach to long-range energetic planning.[14]

Problems

II.1. Prove that in a multi-phase system at equilibrium, the chemical potential of each component is uniform.

Hints: The chemical potential μ_i of component i is the partial molar free energy of i, i.e., an *exergy* term. Calculate the useful work produced by interchanging one mol of i between two phases. Then apply Postulate 1 (Chap. I).

II.2. 1 mol of a monatomic ideal gas is compressed at 573 K from 1 to 10 atm.
a) Assuming a thermostat at 573 K, 1 atm, is available, calculate the minimum useful work necessary for this compression.
b) Calculate the molar exergies of the gas, $\bar{A}'$ and $\bar{A}''$ in the initial and final states, respectively, with reference to a reservoir at 298 K, 1 atm.
c) Show that the difference $\bar{A}'' - \bar{A}'$, calculated from the results of part b) is equal to the maximum useful work which can be obtained from the monothermal expansion of the gas from 10 to 1 atm (both states at 573 K) if only one reservoir, which is at 298 K, 1 atm, is available.

Hint: The molar specific heats at constant pressure and volume are $(5/2)R$ and $(3/2)R$, respectively, R being the universal gas constant [8.314 W s mol^{-1} (K)$^{-1}$].

II.3. A closed system, A, at temperature T_A and pressure p_A undergoes an irreversible change yielding the work W_A, and a heat flow, Q, takes place to the environment, which is at the lower temperature T_0 and pressure p_0. The temperature change of A is negligible.
a) Show that $\Delta_i A_A = -T_0 \Delta_i S_A$, where $\Delta_i A_A$ and $\Delta_i S_A$ are exergy change and entropy creation in A, respectively.
b) Derive an expression for the exergy change of the *global* system (A + reservoir), ΔA_{global}, in terms of the temperatures T_A and T_0, the heat transferred, Q, and the internal entropy change of the system, $\Delta_i S_A$.

II.4. A well-insulated steam turbine is fed by 1000 mol s^{-1} steam. The thermodynamic properties (e. g., molar enthalpy, $\bar{H}_i$, molar entropy $\bar{S}_i$, etc.) of both the incoming steam and the exiting steam-water mixture can be found in tables.

---

14 See, for instance, Henderson 1978, and Odum 1978.

a) Express the power of the turbine, $\mathscr{W}$, in terms of such readily available parameters.

b) A turbine is *perfect* when the rate of power production equals the difference between the incoming and outgoing exergy flow. Express this criterion in terms of the *entropies* of the incoming and outgoing flows of the working fluid.

II.5. Calculate the molar exergy, $\bar{A}$, of superheated water vapor at 644 K, 81.6 atm, with respect to water reservoirs at (a) 25 °C, 1 atm, and (b) 10 °C and 50 atm. The following data from steam tables may be used [15] (Faxen 1953).

Temperature		Pressure	Specific volume	Specific enthalpy	Specific entropy
[°C]	[K]	[bar]	$[dm^3 kg^{-1} = cm^3 g^{-1}]$	$[J\ g^{-1}]$	$[J\ g^{-1}\ K^{-1}]$
370.7	644.0	82.66 = 81.6 atm	30.65	3049	6.212
25	298.3	0.03166[a]	1.003	104.8	0.3671
10	283.3	0.01227[a]	1.004	42.0	0.1511

[a] Pressure of saturated vapor.
The first line in this table refers to vapor, the second and third lines to liquid water.

Selected Literature

(Each section of this list is in reverse chronological order)

Texts and Articles Dealing Entirely with Exergetic Analysis

Kaiser V (1981) Exergy – optimizing every kernel of energy in a system. Chem Eng 88: (No 4) 62
Moran MJ (1981) Availability analysis; a guide to efficient energy use. Prentice-Hall, Englewood Cliffs, New Jersey
Ahern AE (1980) The exergy method for energy systems analysis. Wiley, New York
Evans RB, Crellin GL, Tribus M (1980) In: Spiegler KS, Laird ADK (eds) Principles of desalination. 2nd edn, Ch 1, Academic Press, New York
Gaggioli RA (1980) Thermodynamics: Second law analysis. Am Chem Soc Symp Ser 122: Washington DC
Sussman MV (1980) Availability (exergy) analysis (a self-instruction manual). Mulliken House, 1361 Massachusetts Ave, Lexington, Mass 02173

15 O. H. Faxen: Thermodynamic Tables in the Metric System for Water and Steam, Nordisk Rotogravyrs Monografserie, Haefte 2, Stockholm, 1953.

Literature on Basic Aspects of the Exergy Function

Soma J, Morris HN (1982) Exergy management. In: Payne FW (ed) Energy, economics, policy and management. (Assoc Energy Eng Atlanta Ga)

Oaki H, Ishida M, Ikawa T (1981) Structured process energy-exergy flow diagram and ideality index for analysis of energy transformation in chemical processes (Part I). J Jap Petrol Inst 24:36

Silver RS (1981) Reflexions sur la puissance chaleurique du feu. Heat Recov Sys 1:205 Despite the title, which was chosen in honor of Carnot's classical work, this article is in the English language. It defines *thermergy*, i. e., the heat flow into a building obtainable from a high-temperature source by entirely reversible processes (see Sect. II.E.2.a).

Van Lier JJC (1978) Bewertung der Energieumwandlung bei der Strom- und/oder Wärme-erzeugung. Brennst Wärme-Kraft 30:475

Haywood RW (1974) See literature list for Chapter I

Baehr HD (1966) See literature list for Chapter I

Bosnjakovic F (1965) Technical thermodynamics. Holt, Rinehart and Winston, New York The last chapter of this text is an elementary introduction to exergetic analysis.

Glansdorff P (1957) Sur la fonction dite d'"exergie" et son emploi en climatisation. Suppl Bull Int Cold Inst, Commiss 3 and 6, Padua, Annexes 1957-2

Rant Z (1956) Exergie, ein neues Wort für technische Arbeitsfähigkeit. Forschg Geb Inge-nieurwes 22:36

Van Lerberghe G, Glansdorff P (1932) Le rendement maximum des machines thermiques. Excerpts Publ Associa Eng Mining School Mons No 42

Stodola A (1910) Dampfturbinen. Springer, Berlin, Heidelberg, New York

Gouy J (1889) J Phys Théor Appl 2nd ser 8:501 An early classical contribution explaining the properties of exergy (énergie utilisable) and predicting its importance in practical energetics.

Some Sociological and Environmental Aspects of Thermoeconomics

Verhagen FC (1982) Sociologists and energy engineers. Energy Eng 79:(No 4) 5 (Assoc Energy Eng Atlanta, Ga)

Henderson H (1978) Creating alternative futures. GP Putnam's Sons, New York

Odum HT (1978) Energy analysis, energy quality and environment. In: Gilliland MW (ed) Symposium No 9 Energy analysis: a new public-policy tool. Am Assoc Advanc Sci, Washington DC

Georgescu-Roegen N (1971) The entropy law and the economic process. Harvard Uni-versity Press, Cambridge Mass

Hubbert MK (1971) Energy resources of the earth. Sci Am 224:(No 3) 61 Sept

Other Relevant Literature

Bejan A (1982) Entropy generation through heat transfer and fluid flow. Wiley-Inter-science, New York

Grant EL, Ireson WG, Leavenworth RS (1976) Principles of engineering economy. 6th edn, Wiley, New York

Perry RH, Chilton CH (eds) (1973) Chemical engineer's handbook. 5th edn, McGraw-Hill, New York. See section on *electrodialysis*

Prigogine I (1967) See literature list for Chapter I

Dugdale I (1965) Direct generation of electricity. Spring KH (ed) Academic Press, London

Silver RS (1962) A review of distillation processes for fresh-water production from the sea. Verlag Chemie, Weinheim/Bergstraße Fed Rep Germany, Dechema Monogr 47:19

Lewis GN, Randall M (1923) See literature list for Chapter I

This classical text stresses the application of the Gibbs function, G, which takes the place of exergy, Λ, when system and reservoir are at the same temperature and pressure.

Lotka AJ (1922) Contributions to the energetics of evolution. Proc Natl Acad Sci (USA) 8:147

Chapter III. Generalized Forces

A. Introduction

This chapter deals with the choice of driving forces for transport processes involving one or more flows, which may be coupled, as happens frequently in power-producing or power-consuming processes. For instance, in a thermoelement, heat flow and electric current are coupled. If there is no flow coupling, few restrictions are placed on the choice of useful driving forces. For instance, the driving force for permeation of a gas through a porous material may be considered as the partial-pressure gradient of the chemical-potential gradient of the gas. When related transport processes take place, however, e. g., diffusion and electromigration of ions in electrolytic systems, it is often useful to reduce all driving forces to a common denominator. It is shown in this chapter that the degree of irreversibility of flow processes, as measured by the rate of exergy decrease (or entropy increase) leads to such a common denominator.

B. Choice of Driving Forces

The common characteristic of all irreversible processes, regardless of their physical nature, is an increase of the entropy of the global system, the rate of entropy increase being related to the concomitant rate of exergy decrease by the Gouy-Stodola equation (II-33). Therefore, it makes sense to take proper notice of this common characteristic when comparing different irreversible processes and their mutual interactions, and to define the driving forces for these processes such that, other things being equal, the driving force is the larger the more exergy the irreversible process destroys (or the more entropy it creates). The definition of the driving forces should also take into account that exergy destruction (and entropy creation) are proportional to the flows which give rise to exergy and entropy changes:

$$_{\mathit{\Delta}}X_k \equiv \frac{\dot{S}_k}{\mathscr{I}_k} = - \frac{\dot{A}_k}{T_0\,\mathscr{I}_k}, \tag{III-1}$$

where $\mathscr{I}_k$ is the flow of k, $\dot{S}_k \equiv (d_iS/dt)_k$ is the rate of entropy creation occurring in the system as a result of the flow of k, which could be a flow of heat, mass etc., and $_AX_k$ is the *generalized driving force*, also called thermodynamic force, conjugated to the flow $\mathscr{I}_k$.

Hence the definition of $_AX_k$: The generalized driving force, $_AX_k$, conjugated to the flow $\mathscr{I}_k$ is the entropy creation due to $\mathscr{I}_k$, per second and unit flow, $\mathscr{I}_k$.

$_AX_k$ is thus first determined in a situation in which no flow other than $\mathscr{I}_k$ takes place,[1] and depends on the details of the choice of the current. For instance, in electrochemical problems, one can choose any ionic flow, or the total electric current, and determine the magnitude of the generalized driving force accordingly, i.e., such that the product of flow and generalized force yields the rate of entropy production in the desired units, e.g., W $(K)^{-1}$. The choice of currents or generalized driving force is not always free, however, and it is often necessary to carefully consider how to define the fluxes, if the rules described in Chap. V are to remain valid (Wei 1966). The simple definition of generalized driving forces is sufficient for the simple one-dimensional phenomena of heat, mass and electric-charge transfer treated in this introductory text, however. The *generalized driving forces*, defined here on the basis of rate of entropy creation in a totally irreversible process are evidently not identical, in dimension or units, to the conventional *Newtonian forces* used in mechanics.

In the following, we shall consider four well-known types of irreversible processes and derive the generalized driving forces from the definition [Eq. (III-1)].

1. Heat Transfer (Fig. III-1A)

Heat flows at the rate $\mathscr{I}_Q$ (W) from a mass at temperature T' to another mass at the lower temperature, T''. The rate of entropy creation, $\dot{S}_Q$ (W K^{-1}), in the global system is the algebraic sum of the rates of entropy change in the two masses. Considering the definition of entropy [Eq. (I-2)],[2] this sum is

$$\dot{S}_Q = \mathscr{I}_Q\left(-\frac{1}{T'} + \frac{1}{T''}\right) = \mathscr{I}_Q\Delta\left(\frac{1}{T}\right). \tag{III-2}$$

---

1 When other flows take place simultaneously, these additional flows can increase or decrease the total entropy creation in the system.

2 Although the heat transfer process described here is irreversible, one can, in a thought experiment, imagine the transfer of heat from and to the masses by reversible processes, i.e., by contacting with many reservoirs which are almost, but not quite, at the same temperatures as the masses T' and T'' in Fig. III-1A. One then calculates, by integration, the entropy changes of the latter masses (caused by the loss and gain, respectively, of Q_{rev}) from the *reversible* transfers in the thought experiment, i.e., the heat transfers between bodies having almost the same temperature.

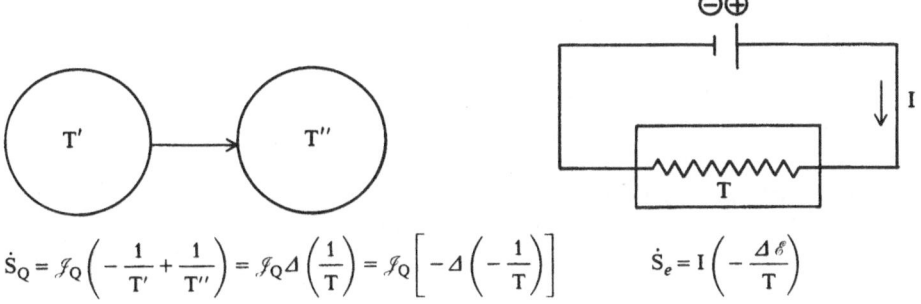

$$\dot{S}_Q = \mathscr{J}_Q\left(-\frac{1}{T'}+\frac{1}{T''}\right) = \mathscr{J}_Q \varDelta\left(\frac{1}{T}\right) = \mathscr{J}_Q\left[-\varDelta\left(-\frac{1}{T}\right)\right] \qquad \dot{S}_e = I\left(-\frac{\varDelta\,\mathscr{E}}{T}\right)$$

A. Heat transfer	B. Isothermal flow of electricity

C. Isothermal ideal–gas interdiffusion   D. Isothermal flow of incompressible fluid

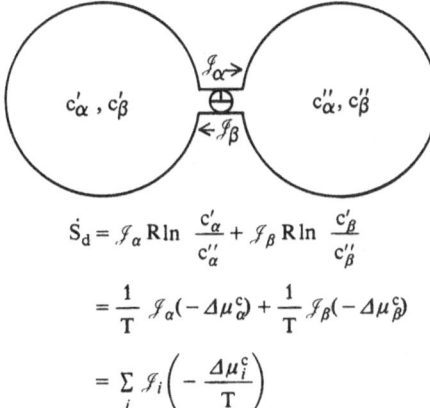

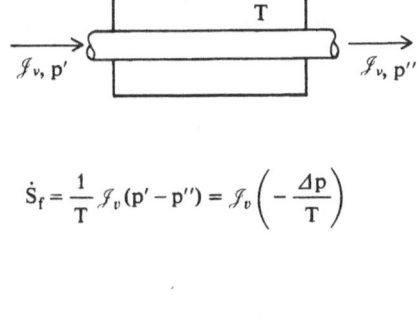

$$\dot{S}_d = \mathscr{J}_\alpha R \ln \frac{c'_\alpha}{c''_\alpha} + \mathscr{J}_\beta R \ln \frac{c'_\beta}{c''_\beta}$$

$$= \frac{1}{T}\,\mathscr{J}_\alpha(-\varDelta\mu^c_\alpha) + \frac{1}{T}\,\mathscr{J}_\beta(-\varDelta\mu^c_\beta)$$

$$= \sum_i \mathscr{J}_i\left(-\frac{\varDelta\mu^c_i}{T}\right)$$

$$\dot{S}_f = \frac{1}{T}\,\mathscr{J}_v(p'-p'') = \mathscr{J}_v\left(-\frac{\varDelta p}{T}\right)$$

Fig. III-1. Four irreversible processes

The operator $\varDelta$ is defined as the parameter (here $1/T$) on the right ($''$) minus the parameter on the left ($'$).[3]

According to the definition (III-1) of the generalized driving force, the force, $_\varDelta X_Q$, conjugated to the heat flow is the rate of entropy creation caused by the heat flow, $\mathscr{J}_Q$, divided by this flow, i.e., [from Eq. (III-2)]:

$$_\varDelta X_Q \equiv \dot{S}_Q/\mathscr{J}_Q = \varDelta\left(\frac{1}{T}\right) = -\varDelta\left(-\frac{1}{T}\right). \tag{III-3}$$

It is seen that the unit of this generalized driving force is $(K)^{-1}$.

---

3 Not the same definition is adopted throughout the literature. In some texts, the sign definition is opposite.

2. Transfer of Electric Charges (Fig. III-1B)

A battery, or any other source of direct current causes the passage of an electric current I (A) through a resistor, $\mathscr{R}$ (Ω), which is placed in a reservoir of temperature T.

The rate of entropy creation, $\dot{S}_e$, is equal to the rate of exergy destruction divided by the reservoir temperature [Eq. (II-30)]. Hence,

$$\dot{S}_e = I\left(-\frac{\Delta\mathscr{E}}{T}\right), \tag{III-4}$$

where $\Delta\mathscr{E}$ is the electric potential difference (V) across the resistor. [Note that when $\Delta\mathscr{E}$ is positive, the electric current is from right to left, i.e., negative, and vice versa, but no matter in which direction the current flows, the product on the right side of Eq. (III-4) is always positive.]

Hence, the generalized driving force is

$$_\Delta X_e \equiv \dot{S}_e/I = -\frac{\Delta\mathscr{E}}{T}. \tag{III-5}$$

The units of this generalized driving force are V $(K)^{-1}$.

3. Isothermal Interdiffusion of Ideal Gases (Fig. III-1C)

The two compartments of the vessel shown in Fig. III-1C contain mixtures of the ideal gases, α and β, respectively. The concentrations c'_α and c''_α are different in the two compartments, but the *total* pressures are equal. (Irreversible) interdiffusion of the two gases starts when the valve separating the two compartments is opened.

When an ideal gas expands isothermally from partial pressure p' to p'', the entropy change is[4]

$$R \ln\left(\frac{p'}{p''}\right) = R \ln\left(\frac{c'}{c''}\right). \tag{III-6}$$

The global entropy creation in this system is the sum of the contributions of the ideal gases, viz.,

---

4 See, for instance, Hatsopoulos and Keenan 1965, p 220.

$$\dot{S}_d = \mathscr{I}_\alpha R \ln \left(\frac{c'_\alpha}{c''_\alpha} \right) + \mathscr{I}_\beta R \ln \left(\frac{c'_\beta}{c''_\beta} \right). \tag{III-7}$$

$\mathscr{I}_\alpha$ and $\mathscr{I}_\beta$ are the flows of gases α and β, respectively (mol s^{-1}).

The concentration ratios can be expressed in terms of differences of chemical potentials by use of the relationship[5]

$$\mu^c = \mu^0 + RT \ln c. \tag{III-8}$$

Here, μ^0 is the standard potential which depends on the temperature, but not on the concentration. Substituting the concentration ratios from Eq. (III-8) into Eq. (III-7), we obtain

$$\dot{S}_d = \frac{1}{T} \mathscr{I}_\alpha(-\Delta\mu_\alpha) + \frac{1}{T} \mathscr{I}_\beta(-\Delta\mu_\beta^c). \tag{III-9}$$

In a multiple-component system, we obtain by reasoning along the same lines:

$$\dot{S}_d = \sum_i \mathscr{I}_i \left(-\frac{\Delta\mu_i^c}{T} \right), \tag{III-10}$$

where subscript i denotes the ith component of the gas mixture.

The generalized driving force for the diffusion of the ith component is obtained by combining Eqs. (III-10) and (III-1):

$$_\Delta X_{d,i} \equiv \dot{S}_{d,i}/\mathscr{I}_i = -\frac{\Delta\mu_i^c}{T}. \tag{III-11}$$

The units of this generalized driving force, X, are W s mol^{-1} K^{-1}.

4. Flow of an Incompressible Fluid (Fig. III-1D)

An incompressible fluid is pumped through a hydraulic resistance, $\mathscr{R}_H$, e.g., a porous sandstone formation, a catalyst bed or a filter, at uniform temperature.

The rate of exergy destruction is $\mathscr{I}_v(p' - p'')$, where $\mathscr{I}_v$ is the volume flow, i.e., the volume being pumped per unit time (cm^3 s^{-1}), and p' and p'' are the pressures at the entrance and exit ends of the hydraulic resistance, respectively.

---

5 See Appendix II. The meaning of the superscript c is that only the variation of the chemical potential with the concentration is considered here; no electrical, elastic, magnetic forces etc. are acting on the system.

The rate of entropy creation can be calculated from the rate of exergy destruction by use of Eq. (II-30):

$$\dot{S}_v = \frac{1}{T}\,\mathscr{J}_v(p' - p'') = \mathscr{J}_v\left(-\frac{\Delta p}{T}\right). \tag{III-12}$$

From the rate of entropy increase, the generalized driving force, conjugated to the volume flow, is calculated from Eq. (III-1):

$$_\Delta X_v \equiv \dot{S}_v/\mathscr{J}_v = -\frac{\Delta p}{T}. \tag{III-13}$$

The units for this generalized driving force, $_\Delta X_v$, are MPa K^{-1} = W s cm^{-3} K^{-1}.[6]

5. Summary of Generalized Driving Forces

Equations (III-5), (III-11) and (III-13) show that the generalized driving forces for the flow of electric charges, diffusion flow and volume flow, respectively, are essentially differences of potential. The generalized driving forces, $-\Delta\mathscr{E}/T$ and $-\Delta p/T$, are similar to the driving forces $-\Delta\mathscr{E}$ and $-\Delta p$ used in electrical engineering and fluid dynamics, respectively, but the generalized driving forces for heat flow, $\Delta(1/T)$, and for diffusion, $-\Delta\mu^c/T$, are not as closely related to the conventional driving forces, $-\Delta T$ (Fourier's law) and $-\Delta c$ (Fick's law) respectively.[7]

In problems involving a single irreversible flow process, it is quite sufficient to use the relevant *conventional* driving force. If different irreversible flow processes are to be compared, however, or if their mutual influence (coupling) is to be studied, it is reasonable to express the flows in terms of the *generalized* driving forces, because the latter are defined in terms of the common characteristic of all irreversible flow processes, viz., the rate of entropy creation (or the rate of exergy destruction).

Table III-1 represents a summary of the notation and the expressions for the generalized driving forces and conjugated flows (or fluxes) discussed in the pre-

---

6 1 MPa = 1 dekabar = 1 W s cm^{-3} = 9.87 atm.

7 Only when the temperature difference, ΔT, across which heat is transferred, is small compared to the absolute temperature, then $\Delta(1/T)$ is approximately proportional to the conventional driving force, $-\Delta T$, because

$$\Delta\left(\frac{1}{T}\right) \equiv \frac{1}{T''} - \frac{1}{T'} = \frac{T' - T''}{T''T'} = -\frac{\Delta T}{T'T''} \approx -\frac{\Delta T}{T_{av}^2}.$$

Table III-1. Generalized driving forces with conjugated flows and fluxes

	Heat transfer	Flow of electric charges	Isothermal diffusion of k	Volume flow
Current (amount s^{-1})	$\mathscr{J}_Q$	I	$\mathscr{J}_k$	$\mathscr{J}_v$
Flux (amount $cm^{-2}\,s^{-1}$)	J_Q	i	J_k	J_v
Generalized driving force, $_\Delta X \equiv \dot{S}/\mathscr{J}$	$_\Delta X_Q = \left(\Delta\frac{1}{T}\right) \approx -\frac{\Delta T}{T_{av}^2}$	$_\Delta X_e = -\frac{\Delta\mathscr{E}}{T}$	$_\Delta X_{d,k} = -\frac{\Delta\mu_k^c}{T}$	$_\Delta X_v = -\frac{\Delta p}{T}$
Differential generalized driving force, $X \equiv \dot{S}/J$	$X_Q = -\frac{1}{T^2}\frac{dT}{dz}$	$X_e = -\frac{1}{T}\frac{d\mathscr{E}}{dz}$	$X_{d,k} = -\frac{1}{T}\frac{d\mu_k^c}{dz}$	$X_v = -\frac{1}{T}\frac{dp}{dz}$
Rayleigh driving force, $_\Delta F \equiv XT$		$_\Delta F_e = -\Delta\mathscr{E}$	$_\Delta F_{d,k} = -\Delta\mu_k^c$	$_\Delta F_v = -\Delta p$
Differential Rayleigh driving force, F		$F_e = -\frac{d\mathscr{E}}{dz}$	$F_{d,k} = -\frac{d\mu_k^c}{dz}$	$F_v = -\frac{dp}{dz}$

ceding. While there exist other force-flow pairs (e. g., in magnetics), those listed in Table III-1 are the only ones used in this introductory text. In addition to the generalized driving forces, $_\Delta X$, the table also lists the *Rayleigh* generalized driving forces, defined as $_\Delta F \equiv {}_\Delta XT$, which are useful in the treatment of systems at uniform temperature. Since for such systems T represents a constant, it can be absorbed into the expressions for the generalized driving forces, $_\Delta F_e$, $_\Delta F_{d,i}$ and $_\Delta F_v$ being used instead of $_\Delta X_e$, $_\Delta X_{d,i}$, and $_\Delta X_v$, respectively, in the flow equations.

This table lists also *differential* generalized driving forces, X_k and F_k, which are useful in continuous systems and are defined as follows:

$$X_k \equiv \frac{\dot{s}_k}{J_k} \tag{III-14}$$

$$F_k \equiv T X_k = T \frac{\dot{s}_k}{J_k}, \tag{III-15}$$

where $\dot{s}$ is the rate of entropy production *per unit volume* ($W \, s \, K^{-1} \, cm^{-3}$) due to the flow of k. While the generalized driving forces $_\Delta X_k$ and $_\Delta F_k$ are calculated from the entropy production, $\dot{S}_k$, in a well-defined region of dimension Δx, Δy, Δz, using the *flow* $\mathcal{J}_k$, [Eq. (III-1)], the *differential* forces X_k and F_k are derived from $\dot{s}_k$ (i. e., the rate of entropy production per cm^{-3}), using the *flux*, J_k, i. e., the flow per unit surface (amount of $k \, cm^{-2} s^{-1}$). The relationship between the two kinds of generalized driving forces is seen by calculating the *differential* force, X, for a volume element $a \Delta z$ of the flow channel:

$$\dot{s} = \lim_{\Delta z \to 0} [\dot{S}/(a \Delta z)] \tag{III-16}$$

$$X \equiv \dot{s}/J = \lim_{\Delta z \to 0} \left(\frac{\dot{S}/(a \Delta z)}{\mathcal{J}/a} \right) = \lim_{\Delta z \to 0} \left(\frac{_\Delta X}{\Delta z} \right). \tag{III-17}$$

It is seen that the *differential* forces contain potential *gradients*, as opposed to the *difference* forces $_\Delta X$ or $_\Delta F$ which contain potential *differences* [see, for example, Eq. (III-11)]. While differential force equations can often be readily integrated along a conduit of varying potential, equations in terms of the difference forces can not; hence, *average* values of the properties in the conduit are often used. The advantages of the differential notation are apparent in the discussion of hyperfiltration (reverse osmosis) in Sect. VIII.C and of thermoelectricity (Chap. X).

C. Comparison of Generalized with Conventional Driving Forces

It is of interest to compare the generalized driving forces, defined by means of a thermodynamic criterion, viz., the rate of entropy production or of exergy destruction [Eq. (III-1)], to the conventional driving forces used by scientists and engineers for the description of the flow phenomena shown in Fig. III-1. The starting point for these considerations is *the continuity equation of the steady state* for the unidimensional flux, J_k, of any conserved quantity, k, e.g., mass or incompressible volume. This equation states that the flux is the product of the (superficial)[8] velocity, u_k, and the concentration, c_k, of k:

$$J_k = u_k c_k .\qquad\text{(III-18)}$$

When the velocity is proportional to the driving force, such as in the electromigration of ions caused by an applied voltage, the ratio $u_k/(\text{driving force})$ is a constant, called the *mobility* of k, and designated as m_k. Introducing the mobility into Eq. (III-18), we obtain

$$J_k = m_k \times c_k \times (\text{driving force on } k) .\qquad\text{(III-19)}$$

To justify the choice of the generalized driving forces of Sect. III.B., it is now shown that they are compatible with the conventional linear laws, e.g., Fick's or Ohm's.

1. Electric Current in an Electrolyte Solution (Uniform Temperature)

The (differential) generalized driving force for this process is listed in Table III-1:

$$F_e = -\frac{d\mathscr{E}}{dz} .\qquad\text{(III-20)}$$

The ion flux, J_e, may be calculated from the electric current density **i** by elementary laws of the physical chemistry of electrolytic solutions (MacInnes 1939):

$$\mathbf{i} = m_{e,k}\underbrace{Z_k\,\mathscr{F}c_k}_{\substack{\text{concentration}\\\text{of the}\\\text{electric charges}}}\left(-\frac{d\mathscr{E}}{dz}\right) = m_{e,k}c_k(-Z_k\,\mathscr{F}\,d\mathscr{E}/dz) ,\qquad\text{(III-21)}$$

---

8 The *superficial* velocity is the (macroscopic) linear velocity of particles k which is superimposed on the statistical velocity fluctuations due to thermal motion.

where Z_k is the valency of the ions k (eq mol^{-1}),[9] c_k the molar concentration of these ions, $\mathscr{F}$ is Faraday's constant (96500 C eq^{-1}) and $m_{e,k}$ is the mobility of these ions.[10]

Since the electrical charge of one mol of ions is $Z_k \mathscr{F}$, the ion flux (mol cm^{-2} s^{-1}) is

$$J_k = i/(Z_k \mathscr{F}) = m_{e,k} c_k \left(-\frac{d\mathscr{E}}{dz} \right) = \frac{m_{e,k}}{Z_k \mathscr{F}} c_k \left(-Z_k \mathscr{F} \frac{d\mathscr{E}}{dz} \right) \tag{III-22}$$

or

$$J_k = m_{e,k} \times c_k \times (-\text{gradient of electric potential}) . \tag{III-23}$$

This flow equation is indeed the conventional law of ion migration.

2. Isothermal Diffusion of Mass k

According to Sect. III.B.3 and Table III-1, the (differential) generalized driving force for this process is

$$F_k = -\left(\frac{d\mu_k^c}{dz} \right)_{p,T} . \tag{III-24}$$

Substituting this (generalized) driving force in Eq. (III-19), we obtain

$$J_k = m_{d,k} c_k \left(-\frac{d\mu_k^c}{dz} \right)_{p,T} . \tag{III-25}$$

If k represents a component of an ideal-gas mixture, or a solute species in an ideal solution, then the chemical potential, μ_k, can be expressed in terms of the concentration, c_k, by Eq. (III-8):

$$J_k = m_{d,k} c_k RT \left(-\frac{d \ln c_k}{dz} \right) = RT\, m_{d,k} \left(-\frac{dc_k}{dz} \right) . \tag{III-26}$$

---

9 For example, $Z_{Na^+} = 1$, $Z_{Ca^{2+}} = 2$, $Z_{Cl^-} = -1$.

10 In some of the literature on the physical chemistry of electrolyte solutions, the expression $m_{e,k}/(Z_k \mathscr{F})$ rather than $m_{e,k}$ is named the mobility of ions k. The definition of the ion mobility, $m_{e,k}$, used in this text is the ion velocity under unit electric-potential gradient, i.e., the ion flux (mol cm^{-2} s^{-1}) divided by the product of the molar concentration and the (negative value of the) electrical potential gradient, in accordance with Eq. (III-23) (MacInnes 1939, p 52).

Since the process is isothermal, the expression $RT\, m_{d,k}$ represents a constant. Hence, Eq. (III-26) is identical with Fick's law of diffusion, the diffusion coefficient (also called diffusivity) of component k being

$$D_k = RT\, m_{d,k}. \tag{III-27}$$

The units of the generalized driving forces $\left(-\dfrac{d\mu_k^c}{dz} \right)_{p,T}$ and $\left(-Z_k \mathscr{F} \dfrac{d\mathscr{E}}{dz} \right)$ in Eqs. (III-24) and (III-22), respectively, are the same, viz., $W\, s\, cm^{-1}\, mol^{-1}$. If it is assumed that the (superficial) velocity of the particles depends only on the magnitude, but not the physical nature, of the generalized driving force, i.e., that k ions neither know nor care whether they are being moved by an electrical or a diffusion force, and that as long as this (generalized) driving force has the same magnitude ($x\, W\, s\, cm^{-1}\, mol^{-1}$), the flux J_k is the same, then we obtain by setting the expressions for J_k in Eqs. (III-22) and (III-25), respectively, equal:

$$m_{d,k} = \frac{m_{e,k}}{Z_k \mathscr{F}}. \tag{III-28}$$

Substituting for $m_{d,k}$ in terms of the ionic diffusion coefficient [Eq. (III-27)], we obtain

$$D_k = RT\, m_{d,k} = RT\, \frac{m_{e,k}}{Z_k \mathscr{F}}. \tag{III-29}$$

This relationship between the diffusion coefficient and the electrical mobility of ions k is known as the *Nernst-Einstein relation.* It is valid in very dilute electrolyte solutions where the basic assumption of equal interactions of the ions with their surroundings is true, irrespective of whether both positive and negative ions move in the same direction (*electrolyte diffusion*), or in opposite directions (*electromigration*).

3. Isothermal Flow of an Incompressible Fluid

According to Table III-1, the differential generalized driving force for this process is

$$F_v = -\frac{dp}{dz}. \tag{III-30}$$

The volume flux ($cm^3\, cm^{-2}\, s^{-1} = cm\, s^{-1}$) which has the same units and equals the linear (superficial) velocity of the fluid is calculated by substituting this driving force in Eq. (III-19):

$$J_v = m_v c_v \left(-\frac{dp}{dz} \right).$$ (III-31)

Inasmuch as in a single-component incompressible fluid, both the concentration, c_v, and the mobility, m_v, are uniform, the product $m_v c_v$ represents a constant, and the flow equation (III-31) is compatible with the conventional law of fluid flow (known as the law of Darcy 1856), describing, for instance, the flow of ground water in permeable strata. This law is of considerable in importance in petroleum technology and hydrology for description of the underground flows of oil and water, respectively.

4. Heat Transfer

From a purely formal viewpoint, one can even try to describe this process by using Eq. (III-19), as was done in Sects. III.C.1 – 3, but since neither the mobility nor the concentration of heat are well defined, this is not attempted here.

According to Table III-1, the generalized driving force for heat flow is

$$_\Delta X_Q = \Delta \left(\frac{1}{T} \right).$$ (III-32)

For temperature differences, $\Delta T \equiv T_2 - T_1$, which are *small compared to the average temperature*, T_{av}, this means that the generalized driving force is proportional to $-\Delta T$, as in Fourier's classical law of heat transfer.

Problems

III.1. A vessel contains a mixture of ideal gases, A and B, respectively, which are not at equilibrium. Although temperature and pressure of the mixture are uniform, the composition is not.

Using Eq. (III-8), express $d\mu_A^c/d\mu_B^c$ in terms of the concentrations of the gases. Relate the result to the Gibbs-Duhem equation [Eq. (VIII-6); see also Katchalsky and Curran 1965, Eq. (V-47)].

III.2. This problem deals with the comparison of diffusion and electrical migration in the following system:
Two large containers are separated by a porous disc of thickness $\Delta z = 0.25$ cm and porosity 0.32.[11] The containers are filled with well-stirred solutions of

---

11 The porosity is defined as the *pore volume* divided by the *total volume* of the disc.

potassium chloride at 25 °C and 1 atm. The volume change in each container is negligible.

a) When the KCl concentrations of the two solutions are 10^{-5} and 10^{-7} mol cm^{-3}, respectively, the steady-state diffusion flux of KCl through the porous disc is $J_{KCl} = 2.97 \times 10^{-10}$ mol $cm^{-2} s^{-1}$. Calculate the diffusion coefficient of KCl in the porous disc, $\bar{D}_{KCl}$, and the resistance factor of the disc. (The resistance factor is the ratio of the diffusion coefficient *in free* solution, D_{KCl} $= 1.90 \times 10^{-5}$ $cm^2 s^{-1}$, to the "*diffusion coefficient in the disc*," $\bar{D}_{KCl} \equiv$ $- J_{KCl} \Delta z / \Delta c_{KCl}$).

Calculate the rate of entropy production per cm^3 of disc, $\dot{s}$, and the rate of exergy disappearance per cm^3, $\dot{\lambda}$. When defining the exergy, use a reference reservoir at 25 °C, 1 atm.

b) After the diffusion measurement, the containers are both filled with solutions of potassium chloride of concentration 10^{-5} mol cm^{-3}. After the KCl concentration in the system has reached uniformity, an electrical potential difference is applied by two suitable electrodes (reversible with respect to Cl^- or K^+) placed in the two containers, respectively. The solutions are well-agitated. Calculate the voltage required to produce a flow of potassium ions of the same magnitude as in a).

c) Calculate the rates of entropy production, $\dot{s}$, and exergy destruction, $\dot{\lambda}$, per cm^3 porous material.

d) Calculate the applied voltage, $\Delta \mathcal{F}$, $\dot{s}$ and $\dot{\lambda}$ for a similar experiment producing the same ion fluxes, using solutions of concentration 10^{-7} mol KCl cm^{-3}.

Hints:

1. Take the mobilities of K^+ and Cl^- equal. Hence, $D_{K^+} = D_{Cl^-} = D_{KCl}$.
2. Remember that in the electromigration experiments, b) to d), the ions K^+ and Cl^- migrate in *opposite directions*, whereas in the diffusion experiment, a), they migrate in the *same direction*.

Selected Literature

On the Choice of Generalized Driving Forces

a) Texts

Haase R (1969) See literature list for Chapter I
Katchalsky A, Curran P (1965) See literature list for Chapter I

b) Article

Wei J (1966) Irreversible thermodynamics in engineering. Ind Eng Chem 58:55
    This article justifies the choice of the generalized driving forces of Table III-1, but shows
    that the simplified method used here is not necessarily valid for all flow processes which
    are outside the scope of this text.

On the Continuity Equation in Steady-State Flow

Jones JB, Hawkins GA (1963) See literature list for Chapter I
  This equation is derived, and discussed in detail, on pp 32 – 34 of this text.
Teorell T (1951) Zur quantitativen Behandlung der Membranpermeabilität. Z Elektrochem
  35:460
  The application of the continuity equation in steady flow [present author's Eq. (3.19)] to
  electrolyte solution and ionic membranes is discussed in detail.

Other References Relevant to the Discussions in Chapter III

Hatsopoulous GN, Keenan JH (1965) See literature list for Chapter I
MacInnes DA (1939) The principles of electrochemistry. 3rd edn, Reinhold, New York
Darcy HPG (1856) "Détermination des lois d'écoulement de l'eau à travers le sable." Les
  fontaines publiques de la Ville de Dijon, Paris, Victor Dalmont. See also the following
  reference:
Hubbert MK (1957) Darcy's law and the field equations of the flow of unterground fluids.
  Bull Associa d'Hydrol Sci #5

Chapter IV. Isothermal Flow Coupling

A. Introduction

The relationships between generalized driving forces and fluxes have been discussed in Chapter III for simple linear cases, when only one force and one flux, which is proportional to this force, are operative. In practice, one often meets more complicated transport processes, however, when more than one force and/or more than one flux are present. In such situations, each flux, J_k, is not only related to its *conjugated* generalized driving force, X_k or F_k, but depends also on all the other forces and/or fluxes. Thus, such fluxes which occur simultaneously are related to each other; they are said to be *coupled*.

In this chapter, the coupling equations are illustrated by a simple example, viz., the internal transfer of masses across an ionic membrane. This example was chosen because (a) it leads to the linear flux equations (including Onsager reciprocity) by elementary physical reasoning, (b) it is important for the analysis of many industrial applications of ionic membranes in which coupled transport occurs (e.g., electrodialysis, battery and fuel-cell technology), and (c) many transport phenomena in living organisms involve coupled flow across ionic membranes. In fact, in vitro experiments with model membranes have contributed greatly to the understanding of passive transport phenomena in living membranes.

The transport phenomena considered in this chapter are *isothermal*. Non-isothermal systems and coupling between heat flow, on the one hand, and mass or electricity flow, on the other, are discussed in Chapters IX and X.

B. Mass Transfer in Ionic Membranes

1. Nature of Ionic Membranes

Two kinds of ionic polymers which can be made into membranes are of major practical importance, viz., cation exchangers and anion exchangers.[1] *Cation-*

---

1 Some authors shorten the term *cat*ion-exchange resin to *cat*ion resin. Note, however, that these materials are polymeric *an*ions. Similarly, *an*ion-exchange resins are polymeric *cat*ions.

exchangers consist of macromolecules with negatively charged ionized groups, e. g., sulfonate groups, $-SO_3^-$, which are attached to the polymer and hence immobile. Their electric charge is counterbalanced by (non-polymeric) positive ions, e. g., Na^+, which can freely migrate within the hydrated polymer, or even exchange position with *cations* in a solution in contact with the cation exchanger. Conversely, the polymeric skeleton of *anion exchangers* contains ionic positive groups, e. g., quarternary amine sites, $-NR_3^+$; their electric charge is counterbalanced by (non-polymeric) *anions* which can migrate through the hydrated polymer and exchange position with anions in solutions contacting the anion-exchanger.

When an electric current is passed through a cation-exchanger, it is carried primarily by positive ions, because almost all *mobile* ions are positive[2]; similarly, anion exchangers are primarily anionic conductors. In a purely phenomenological sense, there exists an analogy with p and n type semiconductors, respectively.

Many different transport phenomena can take place when an ionic membrane separates two electrolyte solutions. The *electric conductance* of the membrane, due to *electromigration* of the ions can be measured. *Self-diffusion* of ions is measured by observing the flow of radioactive ions (or other isotopic tracer ions) between solutions of equal concentration separated by the membrane. If the solution concentrations are different, simultaneous *counterion and coion diffusion* takes place. In general, an electric potential difference then develops across the membrane. This *membrane potential* can be measured with identical reference electrodes (e. g., calomel electrodes equipped with saturated KCl bridges) placed in the two solutions, respectively.[3] When a pressure difference is applied across the membrane, *streaming potentials* develop across the membrane and *hyperfiltration* (reverse osmosis) can take place, i. e., the membrane acts as a (molecular) salt filter. The methods described in this chapter lead to the understanding of the relationships between these different phenomena.

2. Summation of Different Generalized Driving Forces

To illustrate these relations, we consider the *simultaneous action* of concentration, electric-potential and pressure gradients in an isothermal membrane, and

---

2 When a cation exchanger is placed into an electrolyte solution, relatively few anions penetrate into the solid exchanger. The thermodynamic reasons for this phenomenon, known as *Donnan effect*, are discussed in the literature on the electrochemistry of ion-exchange materials (for summaries, see, for instance, Helfferich 1962, Moore 1968). The majority of the mobile ions (*counter*ions) in a cation exchanger are thus cations, while only relatively small amounts of mobile ions of opposite charge (coions) can enter the solid. In analogy, the counterions in an anion exchanger are anions; only small amounts of positive coions can penetrate into this solid.

3 For a review of the techniques and interpretation of such measurements in vitro, see, for instance, Moore (1968); for in vivo measurements, Cole (1972) and Tasaki (1968).

express the fluxes all as mol cm^{-2} s^{-1}. Thus, instead of the electric current density, i (Table III-1), caused by the flow of species γ, we express the transport in terms of the molar flux of this species, γ. Since $i_\gamma = Z_\gamma J_\gamma$ (Z_γ being the valency of the ion γ, positive for cations, negative for anions and zero for water, and $\mathscr{F}$ is Faraday's constant, $0.965 \times 10^5\ C\ Eq^{-1}$), it follows from Eq. (III-15) that the generalized electric driving force conjugated to J_γ is $F_{e,\gamma} = -Z_\gamma(d\mathscr{E}/dz)$. The molar flux of γ due to a pressure gradient is obtained from the volume flux by dividing it by the partial molar volume of γ, $\bar{V}_\gamma$: $J_{p,\gamma} = J_v/\bar{V}_\gamma$. The generalized pressure driving force, conjugated to $J_{p,\gamma}$ is obtained from Eq. (III-15) as $F_{p,\gamma} = -\bar{V}_\gamma(dp/dz)$.

Keeping in mind that all forces, F_γ, must be expressed in terms of the same units before they can be added,[4] the *total* (Rayleigh) generalized driving force is

$$F_\gamma = -\left(Z_\gamma\,\mathscr{F}\,\frac{d\mathscr{E}}{dz} + \bar{V}_\gamma\frac{dp}{dz} + \frac{d\mu_\gamma^c}{dz}\right)$$

$$= -\left(Z_\gamma\,\mathscr{F}\,\frac{d\mathscr{E}}{dz} + \frac{d\mu_\gamma}{dz}\right) \tag{IV-1}$$

$$= -\frac{d\tilde{\mu}_\gamma}{dz},$$

where μ_γ is the chemical potential (of species γ) which depends on concentration and pressure, while $\tilde{\mu}_\gamma$ is named the *total potential*. For each species in the membrane, the gradient of its total potential represents the generalized driving force for transport. This expression is a generalization of the differential Rayleigh driving force for isothermal diffusion of uncharged particles at constant total pressure [Eq. (III-11); Table III-1].

Using this generalized *driving force*, we can express the *flux* of species γ by means of the steady-state continuity equation, (III-19):

$$J_\gamma = m_\gamma c_\gamma F_\gamma = \frac{D_\gamma}{RT}c_\gamma\left[-\left(Z_\gamma\,\mathscr{F}\,\frac{d\mathscr{E}}{dz} + \bar{V}_\gamma\frac{dp}{dz} + \frac{d\mu_\gamma^c}{dz}\right)\right]. \tag{IV-2}$$

Here, m_γ is the mobility of species γ, which is operationally defined from Eqs. (III-28) and (III-29).

The flux equation (IV-2) may be rewritten in shorter from by defining the *conductance coefficient*, $L_{\gamma\gamma}$ (Onsager 1931):

---

4 This implies expression of the pressure, p, in $W\,s\,cm^{-3}$ = megapascal (1 MPa = 9,87 atm).

$$m_\gamma c_\gamma = \frac{D_\gamma}{RT} c_\gamma \equiv L_{\gamma\gamma}. \qquad\qquad (IV\text{-}3)$$

Hence:

$$J_\gamma = L_{\gamma\gamma} F_\gamma. \qquad\qquad (IV\text{-}4)$$

Thus, the flux of any species γ is the product of the generalized driving force acting on this species and the conductance coefficient, $L_{\gamma\gamma}$, which is concentration-dependent, as can be seen from its definition [Eq. (IV-3)]. This coefficient may be considered as constant in homogeneous, isothermal conductors, e. g., the metallic wires or semiconductors forming a thermoelement in which the concentration of the charge carriers is uniform. When a single flux, J_γ, takes place in such media, it is *proportional* to the conjugated generalized force, F_γ.

C. Linear Phenomenological Relations Between Fluxes and Generalized Driving Forces

1. Interaction Between Simultaneous Fluxes

Equation (IV-4) expresses the flux, J_γ, caused by the generalized driving force F_γ. When other forces and/or other fluxes are present, however, the value of this flux may be different, because the flux of any given species, γ, is not only influenced by its conjugated force, F_γ. It is necessary to take the influence of other (non-conjugated) forces into account also. *All* the forces and fluxes present can contribute to the total entropy production $\dot{S} = \sum_i J_i X_i$ [Eq. (IV-1)] due to the irreversible processes taking place in the system. The mutual influence between such simultaneous irreversible processes is called *coupling*.

Thus, particle flow can take place even in the absence of a conjugated force: for instance, in ionic membranes, the application of an electric field causes not only conjugated electromigration of the *ions*, but also the flow of water (*electroosmotic flow*) although the water molecules carry no net electric charge and are merely dragged along by the counterions.[5]

From a purely formal viewpoint, the influence of non-conjugated forces, $F_j (j \neq i)$ on the flux J_i can be described by a linear expansion of the following kind:

---

5 Electroosmosis is discussed in greater detail in Sects. IV.C.2.d. and IV.D.

$$
\left.
\begin{aligned}
J_1 &= L_{11}F_1 + L_{12}F_2 + L_{13}F_3 + \ldots + L_{1i}F_i \\
J_2 &= L_{21}F_1 + L_{22}F_2 + L_{23}F_3 + \ldots + L_{2i}F_i \\
J_3 &= L_{31}F_1 + L_{32}F_2 + L_{33}F_3 + \ldots + L_{3i}F_i \\
&\vdots \qquad \vdots \qquad \vdots \qquad \vdots \qquad \vdots \\
J_i &= L_{i1}F_1 + L_{i2}F_2 + L_{i3}F_3 + \ldots + L_{ii}F_i
\end{aligned}
\right\}
\qquad \text{(IV-5)}
$$

In principle, a linear expansion of this kind can be expected to be a reasonably good approximation only for a small range of driving forces, F_i, i.e., close to equilibrium when all these driving forces are infinitesimal. For many mass-transport phenomena, however, these linear approximations are often valid quite far from equilibirum, i.e., when the driving forces are not infinitesimal. The validity of linear laws [such as the conventional law of ion migration, Fick's law of diffusion, and Darcy's flow law, which are compatible with Eq. (IV-4) as shown in Sect. III.C.], over a broad range of driving forces indicates that the flux equations (IV-5) are useful for systems with finite driving forces and fluxes. For the isothermal membrane system considered here, this conclusion is illustrated in Sect. IV.C.2.a. by means of the friction model, which is based on Newtonian physics. On the other hand, the range of validity of linear expansions of the type of Eqs. (IV-5) for coupled chemical reactions (Chap. VI) is usually quite small, because chemical reactions in general consist of a complex sequence of elementary steps. Hence, for chemical reactions, linear expansions of the type of Eqs. (IV-5) are valid only quite close to equilibrium (Onsager 1931). This is also true for a variety of other coupled processes.

Instead of explicitly expressing the fluxes as functions of the driving forces [Eqs. (IV-5)], it is often more convenient to express the generalized driving forces as functions of the fluxes, by inverting the matrix of the conductance coefficients in Eqs. (IV-5). The result is

$$
\left.
\begin{aligned}
F_1 &= R_{11}J_1 + R_{12}J_2 + R_{13}J_3 + \ldots + R_{1i}J_i \\
F_2 &= R_{21}J_1 + R_{22}J_2 + R_{23}J_3 + \ldots + R_{2i}J_i \\
F_3 &= R_{31}J_1 + R_{32}J_2 + R_{33}J_3 + \ldots + R_{3i}J_i \\
&\vdots \qquad \vdots \qquad \vdots \qquad \vdots \qquad \vdots \\
F_i &= R_{i1}J_1 + R_{i2}J_2 + R_{i3}J_3 + \ldots + R_{ii}J_i
\end{aligned}
\right\}
\qquad \text{(IV-6)}
$$

where the parameters R_{ij} are called *resistance coefficients* in analogy to Ohm's law.

Both the conductance coefficients L_{ij} and the resistance coefficients, $R_{ij}(i \neq j)$ are quantitative expressions for coupling. To relate them to a physical model, and to illustrate the reciprocity relations, Sect. IV.C.2. introduces the friction model of steady-state transport in viscous media.

2. Flow Coupling in Membranes

a) Friction Model

A unit volume of membrane material, m, including the mobile positive counterions (s) and mobile water (w) is represented in Fig. IV-1. The concentration of
(mobile) negative coions is negligible. The temperature is constant. A generalized
driving force, F_s (W s mol^{-1} cm^{-1}), e.g., an electric-potential gradient, causes
migration of the counterions. In the *steady state*, the driving force is counter-
acted by friction forces between the counterions and the membrane, F_{sm}, and
between the counterions and water, F_{sw}, respectively (all forces refer to 1 mol of
counterions). In this state, no *net* force acts of the counterions. Hence, their
superficial velocity, u_s, is constant.[6]

The size of the hydrated counterions is larger than that of the water molecules. Because the major purpose of the model is the derivation of *phenomenological* flux equations of the type (IV-5) or (IV-6), rather than the determination
of exact geometrical parameters (such as the shape of the counterions), it is permissible to use the equation of viscous friction in a continuum, as is done frequently in the study of diffusion and electromigration of ions in solutions. In
particular, it is assumed that the interaction (friction) force between particles i
and j, F_{ij}, is proportional to the difference between the velocities u_i and u_j, respectively. The factor of proportion is called the friction coefficient f_{ij} (W s^2
cm^{-2} mol^{-1}):

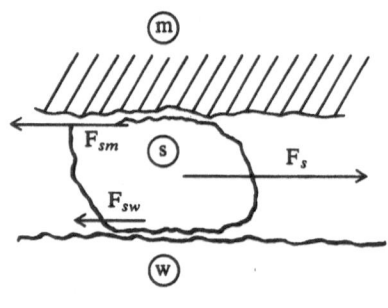

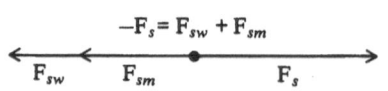

Fig. IV-1. Friction model for mass transfer in
membranes. The figure represents unit volume of
an ion-exchange membrane equilibrated with a
very dilute aqueous electrolyte solution. s counterions; w water molecules; m solid membrane
polymer

---

6 This state of balance of forces is sometimes named *mechanical equilibrium*. Note that in
this state, irreversible flow processes take place, and entropy is being created, whereas in
a system at genuine *thermodynamic equilibrium* (as defined in Sect. I.C.), entropy is not
created.

$$F_{ij} = - f_{ij}(u_i - u_j) .\tag{IV-7}$$

Since the friction force, F_{ij}, refers to the interaction between one mol i and all the particles j present in the unit volume described by Fig. IV-1, so does the friction coefficient, f_{ij}, which characterizes the interactions between one mol i and all the particles j in the unit volume of membrane schematically shown in Fig. IV-1. Since the interaction force between particles i and j is likely to be proportional to the number of $i-j$ collisions, and hence to the concentrations of both i and j, it is reasonable to define a *mutual friction coefficient*, ζ_{ij}, normalized with respect to the concentration of species j, i.e., a coefficient which expresses the interaction between one mol of i and one mol of j:

$$\zeta_{ij} \equiv \frac{f_{ij}}{\bar{c}_j} ,\tag{IV-8}$$

which is symmetrical with respect to particles i and j:

$$\zeta_{ij} = \zeta_{ji} = \frac{f_{ji}}{\bar{c}_i} .\tag{IV-9}$$

The units of this coefficient are $W\,s^2\,cm\,mol^{-2}$. The overbar indicates concentrations *in the membrane*, i.e., the units of $\bar{c}$ are $mol/(cm^3$ membrane volume).

The force balance for the counterions, s, is (Fig. IV-1):

$$F_s = - (F_{sw} + F_{sm}) = f_{sw}(u_s - u_w) + f_{sm}(u_s - u_m)\tag{IV-10}$$

and for one mol water the force balance is:

$$F_w = f_{ws}(u_w - u_s) + f_{wm}(u_w - u_m) .\tag{IV-11}$$

The friction coefficient, f_{ws}, in this equation can be expressed in terms of f_{sw} by means of Eqs. (IV-8) and (IV-9):

$$f_{ws} = \bar{c}_s \zeta_{ws} = \bar{c}_s \zeta_{sw} = \frac{\bar{c}_s}{\bar{c}_w} f_{sw} .\tag{IV-12}$$

Substituting this equation in (IV-11), we obtain:[7]

---

7 The choice of $u_m = 0$ amounts to measuring all (superficial) velocities, u_i, with respect to the fixed membrane. The literature often describes particle motion in terms of other frames of reference, e.g., with respect to the center of gravity or the velocity of the solvent. In these systems of coordinates, the velocity is often not uniform even in the steady state, i.e., even in this state, u_i, varies with z. Conversion of flux formulae expressed in one system of coordinates into another is described by Miller (1960) and Haase (1969).

$$F_w = \frac{\bar{c}_s}{\bar{c}_w} f_{sw}(u_w - u_s) + f_{wm}u_w. \tag{IV-13}$$

To express the driving forces, F, in terms of fluxes, instead of (superficial) velocities, we use Eq. (III-18), $J = u/\bar{c}$, and introduce the fluxes into Eqs. (IV-10) and (IV-13):

$$F_s = \frac{f_{sw} + f_{sm}}{\bar{c}_s} J_s - \frac{f_{sw}}{\bar{c}_w} J_w \tag{IV-14}$$

$$F_w = - \frac{f_{sw}}{\bar{c}_w} J_s + \frac{(\bar{c}_s/\bar{c}_w) f_{sw} + f_{wm}}{\bar{c}_w} J_w. \tag{IV-15}$$

Instead of the friction coefficients, f_{ij}, the coefficients ζ_{ij} (symmetrical with respect to i and j) may be introduced from Eq. (IV-8):

$$F_s = - \frac{1}{\bar{c}_s}(\bar{c}_w\zeta_{sw} + \bar{c}_m\zeta_{sm})J_s - \zeta_{sw}J_w \tag{IV-16}$$

$$F_w = - \zeta_{sw}J_s + \frac{1}{\bar{c}_w}(\bar{c}_s\zeta_{ws} + \bar{c}_m\zeta_{wm})J_w. \tag{IV-17}$$

It is seen that the friction model leads to equations of the type (IV-6) and that the resistance coefficients, R, are functions of the concentrations of the counterions and of water in the membrane.[8]

In all considerations of the friction model summarized in Eqs. (IV-14) to (IV-17), it has been assumed that the concentration of the coions can be neglected compared to the concentration of the counterions. This is a reasonable assumption when the molar concentration of fixed charges in the membrane is much larger than the molar electrolyte concentration in the solutions contacting the membrane (*Donnan effect*). If this is not the case, the concentration of coions cannot be neglected. It is very important to keep in mind that the resistance coefficients such as $R_{ss} = -(1/\bar{c}_s)(\bar{c}_w\zeta_{sw} + \bar{c}_m\zeta_{sm})$ [Eq. (IV-16)] depend strongly on the solution concentrations, and so do the conductance coefficients, L_{ij}. This dependence of R_{ij} and L_{ij} is much more pronounced than the dependence, on solution concentration, of ionic diffusion coefficients and ionic mobilities.

---

8 The membrane concentration, $\bar{c}_m$, and the friction coefficients, ζ_{im}, are defined with respect to the molar concentration of fixed-charge groups (often called *ion-exchange capacity*) of the membrane, but other units of concentration may be used, provided that the products $\bar{c}_{im}\zeta_{im}$ have the units $W\ s^2\ cm^{-2}\ (mol\ i)^{-1}$.

b) Reciprocity of Coupling Coefficients

Comparison of Eqs. (IV-6) with (IV-16) and (IV-17) demonstrates that the resistance coefficients characterizing the coupling of fluxes J_i and J_j, namely R_{ij} and R_{ji} are equal. Because of the equality of the resistance coefficients, R, in Eqs. (IV-6), the corresponding conductance coefficients, L, in Eqs. (IV-5) are also equal, because the matrix of (IV-6) is obtained simply by inversion of the symmetric matrix of (IV-5) (as shown in texts of algebra)

$$R_{ij} = R_{ji} \tag{IV-18}$$

$$L_{ij} = L_{ji}. \tag{IV-19}$$

In the case of isothermal mass transport considered here, the relations (IV-18) or (IV-19), called *reciprocity relations*, result from the mechanical law of numerical equality of action and reaction, as expressed by the equality of the mutual friction coefficients in Eq. (IV-9). Onsager (1931) has demonstrated the generality of these reciprocity relations for many different flow phenomena, including those which cannot be described by the simple friction model, provided the fluxes and forces are properly chosen and conjugated to each other.[9]

The reciprocity relations (IV-18) and (IV-19) both express the symmetry of flow coupling. We can readily demonstrate the equality of the conductance coefficients, $L_{ij} = L_{ji}$ from the equality of the resistance coefficients, $R_{ij} = R_{ji}$, specifically for the case considered here by solving the two simultaneous equations (IV-14) and (IV-15) in terms of the fluxes J_s and J_w:

$$J_s = \frac{\bar{c}_s}{d'} (\bar{c}_s f_{sw} + \bar{c}_w f_{wm}) F_s + \frac{\bar{c}_s \bar{c}_w f_{sw}}{d'} F_w \tag{IV-20}$$

$$J_w = \frac{\bar{c}_s \bar{c}_w f_{sw}}{d'} F_s + \frac{\bar{c}_w^2 (f_{sw} + f_{sm})}{d'} F_w, \tag{IV-21}$$

where:

$$d' \equiv \bar{c}_s f_{sw} f_{sm} + \bar{c}_w (f_{sw} + f_{sm}) f_{wm}. \tag{IV-22}$$

---

9 For coupled-flow equations, reciprocity of transport coefficients holds only when driving forces are conjugated to fluxes, as explained in Chap. III [Eq. (III-1) and Table III-1]. For instance, in a common description for *diffusion in multicomponent systems* (Miller 1960), this is (purposely) not done, and hence the diffusion coupling terms, D_{ij} and D_{ji}, respectively, are not equal:

$$\left. \begin{array}{l} J_i = D_{ii}(-dc_i/dz) + D_{ij}(-dc_j/dz) \\ J_j = D_{ji}(-dc_i/dz) + D_{jj}(-dc_j/dz) \end{array} \right\} \quad D_{ij} \neq D_{ji}.$$

(*i* and *j* refer to two salts in aqueous solution, diffusing simultaneously.)

It is seen that Eqs. (IV-20), (IV-21) are indeed of the type of Eqs. (IV-5):

$$J_s = L_{ss}F_s + L_{sw}F_w \tag{IV-23}$$

$$J_w = L_{ws}F_s + L_{ww}F_w . \tag{IV-24}$$

From Eqs. (IV-20), (IV-21) it is seen that the conductance coefficients, L_{sw} and L_{ws} are indeed equal, as are the resistance coefficients, $R_{sw} = R_{ws} = -f_{sw}/\bar{c}_w$ in Eqs. (IV-14), (IV-15).

Substitution of the friction coefficients, f_{ij} in Eqs. (IV-21), (IV-22) in terms of the mutual friction coefficients, ζ_{ij} [Eq. (IV-8)] leads to the following expressions for the conductance coefficients:

$$L_{ss} = \frac{\bar{c}_s(\bar{c}_s\zeta_{sw} + \bar{c}_m\zeta_{wm})}{\bar{c}_m d''} \tag{IV-25}$$

$$L_{sw} = L_{ws} = \frac{\bar{c}_s\bar{c}_w\zeta_{sw}}{\bar{c}_m d''} \tag{IV-26}$$

$$L_{ww} = \frac{\bar{c}_w(\bar{c}_w\zeta_{sw} + \bar{c}_m\zeta_{sm})}{\bar{c}_m d''} \tag{IV-27}$$

where

$$d'' = \bar{c}_s\zeta_{sw}\zeta_{sm} + \bar{c}_w\zeta_{sw}\zeta_{wm} + \bar{c}_m\zeta_{sm}\zeta_{wm} . \tag{IV-28}$$

c) Rationale for Use of Friction Coefficients

While not all irreversible processes involve friction,[10] the description of mass transfer in solutions, melts, diaphragms and membranes, in terms of friction coefficients, often proves useful. The basic observation which leads to this description is the approximate constancy (with respect to concentration changes) of *diffusion coefficients*, D_k, defined by Fick's law:

$$J_k = D_k(-dc_k/dz) . \tag{IV-29}$$

At a given temperature, this fact implies also approximately constant *friction coefficients*, since diffusion and friction coefficients are related by a simple law, which is strictly valid only for Brownian motion of colloids (Einstein 1905), but is approximately valid also for the diffusion of many other species in dilute solutions:

$$D_k = \frac{RT}{f_{kw}} . \tag{IV-30}$$

---

10 For instance, no mechanical friction is involved in the transfer of heat from high to low temperature.

On the other hand, if flux-force relations are expressed in terms of conductance coefficients, L_{ij}, one obtains from Eqs. (III-25) and (III-27):

$$J_k = L_{kk}F_k = \frac{D_k}{RT}\,\bar{c}_k\left(-\frac{d\mu_k^c}{dz}\right),$$ (IV-31)

where

$$L_{kk} = \frac{1}{f_{kw}}\,\bar{c}_k.$$ (IV-32)

In dilute solutions, where f_{kw} is almost constant, L_{kk} is approximately proportional to the concentration. In concentrated solutions and in membranes, the friction coefficients vary to some extent with the concentration, but sometimes the use of friction coefficients sheds more light on the mechanism of flux interaction than the use of the conductance coefficients, which combine many different friction and concentration effects, as seen from Eqs. (IV-20), (IV-21).

Systematic studies of friction coefficients have shown that in cation-exchange membranes, the friction cation ↔ polymer is considerably higher than the frictions cation ↔ water or water ↔ polymer. This is an indication of the close proximity of the cations to the negatively charged polymer. On the other hand, studies of the permeation of sugars through cellophane have shown that the sugar ↔ membrane friction coefficient, which increases with molecular size of the sugar and decreasing temperatures, is smaller than the sugar ↔ water friction coefficient. This indicates that in this system there is no tendency of the solute molecules to migrate in the vicinity of the polymer backbone. Moreover, the water ↔ membrane friction coefficients were found to be of the same order as in commercial ion-exchange polymers. All these findings are in agreement with the friction model (Fig. IV-1).

d) Electroosmosis: Comparison of Friction-Coefficient to Helmholtz-Theory Expressions

The consideration of *electroosmosis*, i.e., the coupled (drag) flow of solvent caused by ion flow in ionic membranes, is of intrinsic scientific interest, but also of practical importance, because electroosmotic pumping of solvent has been investigated with a view to energy conversion. Coupling efficiencies are discussed in general macroscopic (phenomenological) terms in Chapter V, but it is worthwhile to consider some friction-model concepts of electroosmosis at this point.

The ratio of water to counterion flux in isothermal *electroosmosis* is obtained when one divides Eq. (IV-21) by (IV-20), or (IV-24) by (IV-23), for $F_w = 0$, because in electroosmosis no generalized driving force is directly applied to water molecules, which are dragged along by the counterions migrating under the influence of an electric driving force, F_s:

$$\left(\frac{J_w}{J_s}\right)_{F_w=0} = \frac{L_{ws}}{L_{ss}} = \frac{\bar{c}_s\bar{c}_w f_{sw}}{\bar{c}_s(\bar{c}_s f_{sw} + \bar{c}_w f_{wm})} = \frac{\bar{c}_w}{\bar{c}_s + \bar{c}_w \dfrac{f_{wm}}{f_{sw}}} \qquad \text{(IV-33)}$$

or, substituting *mutual* friction coefficients, ζ_{ij}, for the friction coefficients, f_{ij}, by means of Eq. (IV-8):

$$\left(\frac{J_w}{J_s}\right)_{F_w=0} = \frac{\bar{c}_w}{\bar{c}_s + \bar{c}_w \dfrac{\zeta_{wm}\bar{c}_m}{\zeta_{ws}\bar{c}_w}} = \frac{1}{\dfrac{\bar{c}_s}{\bar{c}_w} + \dfrac{\bar{c}_m\zeta_{wm}}{\bar{c}_w\zeta_{ws}}}. \qquad \text{(IV-34)}$$

It is seen that, in general, the electroosmotic water transport is lower than the concentration ratio $\bar{c}_w/\bar{c}_s$ in the membrane. For instance, in many cation-exchange membranes with sodium (or other alkali-metal) counterions, the *electroosmotic velocity ratio* $(J_w/J_s)/(\bar{c}_w/\bar{c}_s) = u_w/u_s$ is approximately 1/2 (Spiegler 1953, 1958, Paterson 1976, Demisch and Pusch 1976).

These experimental findings mean that the linear (superficial) velocity of water in ionic membranes with sodium counterions is substantially lower than the counterion velocity, whereas classical electrokinetic theory assumes equality of these two velocities, although Helmholtz himself pointed out that the equations based on this assumption are only applicable to media with relatively large characteristic pore dimensions. This is the case for glass capillaries, in which sodium ions at the glass/solvent interface are the counterions, that can migrate in an applied electric field and drag the solvent in the interior of the capillary along. In solvent-swollen ion-exchange membranes which are solid polyelectrolytes, the distances between polymer chains are generally less than 2×10^{-2} μm (200 Å), however. Hence the pores are neither of atomic dimensions, nor macroscopic. Because of the colloidal nature of the membranes, macroscopic models frequently do not reflect their transport behavior. The simple friction model is applicable to membranes, despite their characteristic colloid dimensions. When used judiciously, it offers a semi-quantitative description of transport phenomena in homogeneous membranes. For non-homogeneous membranes, such as layered (sandwich-type) or mosaic (patch-type) membranes, however, the equations have to be suitably modified so as to avoid results which cannot be interpreted in terms of the simple model (Fig. IV-1), such as negative friction coefficients. Moreover, it is important to keep in mind that the basic friction equations (IV-10), (IV-11) and the conclusions therefrom cover viscous (laminar) flow processes, i. e., those in which flux and generalized driving force are proportional to each other. Turbulent-flow phenomena, which are frequently of major importance in transfer processes, are not described by these equations.

D. Electrokinetic Flow Equations

It is useful to apply the flow equations (IV-23) and (IV-24) to systems of membranes and solutions in which only electric and/or pressure forces act across the membrane.[11] The study of these phenomena is called *electrokinetics*. Electroosmosis (Sect. IV.C.2.d) is an electrokinetic phenomenon.

Consider a separator containing fixed (immobile) negative charges, mobile counterions (Na^+), and water (w). The system is isothermal. In this case, Eqs. (IV-24) and (IV-23), respectively, are

$$J_w = L_{ww}F_w + L_{wNa}F_{Na} \tag{IV-35}$$

$$J_{Na} = L_{Naw}F_w + L_{NaNa}F_{Na}, \tag{IV-36}$$

where

$$L_{Naw} = L_{wNa} \tag{IV-37}$$

as shown by Eqs. (IV-20) and (IV-21).

The Rayleigh generalized driving forces are

$$F_w = -\bar{V}_w \Delta p \tag{IV-38}$$

$$F_{Na} = -\bar{V}_{Na} \Delta p - Z_{Na} \mathscr{F} \Delta \mathscr{E} \tag{IV-39}$$

as shown in Sect. IV.B.2. $\bar{V}$ and Z are the partial molal volume and the valency of the particles, respectively, ($Z_{Na} = 1$).

Substituting these generalized driving forces in Eqs. (IV-35) and (IV-36), we obtain

$$-J_w = (L_{ww}\bar{V}_w + L_{wNa}\bar{V}_{Na})\Delta p + L_{wNa}Z_{Na}\mathscr{F}\Delta\mathscr{E} \tag{IV-40}$$

$$-J_{Na} = (L_{Naw}\bar{V}_w + L_{NaNa}\bar{V}_{Na})\Delta p + L_{NaNa}Z_{Na}\mathscr{F}\Delta\mathscr{E}. \tag{IV-41}$$

---

11 The term membrane is used here in a general sense; it applies to any separator which is an ionic conductor, placed between solutions. For instance, porous glass disks are included.

Hence, the total volume flux and the current density, respectively, are

$$- J_v = \bar{V}_w J_w + \bar{V}_{Na} J_{Na} = (L_{ww} \bar{V}_w^2 + 2 L_{wNa} \bar{V}_{Na} \bar{V}_w$$
$$+ L_{NaNa} \bar{V}_{Na}^2) \Delta p + Z_{Na} \mathscr{F} (L_{wNa} \bar{V}_w + L_{NaNa} \bar{V}_{Na}) \Delta \mathscr{E} \qquad \text{(IV-42)}$$

$$- i = - Z_{Na} \mathscr{F} J_{Na} = Z_{Na} \mathscr{F} (L_{Naw} \bar{V}_w + L_{NaNa} \bar{V}_{Na}) \Delta p + Z_{Na}^2 \mathscr{F}^2 L_{NaNa} \Delta \mathscr{E}.$$
$$\text{(IV-43)}$$

We rewrite these equations in the form of Eqs. (IV-5):

$$J_v = L_{vv} (- \Delta p) + L_{vi} (- \Delta \mathscr{E}) \qquad \text{(IV-44)}$$

$$i = L_{iv} (- \Delta p) + L_{ii} (- \Delta \mathscr{E}). \qquad \text{(IV-45)}$$

Comparing these equations with Eqs. (IV-42) and (IV-43), it is seen that the reciprocity relation $L_{iv} = L_{vi}$ is satisfied.

By solving the simultaneous equations (IV-44) and (IV-45) for $- \Delta p$ and $- \Delta \mathscr{E}$, we obtain equations of the type of (IV-6):

$$- \Delta p = R_{vv} J_v + R_{vi} i \qquad \text{(IV-46)}$$

$$- \Delta \mathscr{E} = R_{iv} J_v + R_{ii} i. \qquad \text{(IV-47)}$$

The matrix inversion preserves reciprocity, i.e., $R_{iv} = R_{vi}$.

Consider the application of pressure to the membrane-solution system shown in Fig. IV-2(a). A potential difference, called *streaming potential*, develops across the membrane, which is measured with a voltmeter of very high impedance ($i \approx 0$). From Eq. (IV-45), we obtain for the *streaming-potential coefficient*:

$$\left(\frac{\Delta \mathscr{E}}{\Delta p} \right)_{i=0} = - \frac{L_{iv}}{L_{ii}}. \qquad \text{(IV-48)}$$

Now consider the application of an electric-potential difference in the absence of a pressure gradient [Fig. IV-2(b)]. The electromigration of ions causes *electroosmotic flow* of solvent, as described in Sect. IV.C.2.d. From the total volume flux, J_v, the *electroosmotic coefficient*, β, can be determined by dividing Eq. (IV-44) by Eq. (IV-45):

$$- \left(\frac{J_v}{i} \right)_{\Delta p = 0} = - \frac{L_{vi}}{L_{ii}} \equiv - \beta. \qquad \text{(IV-49)}$$

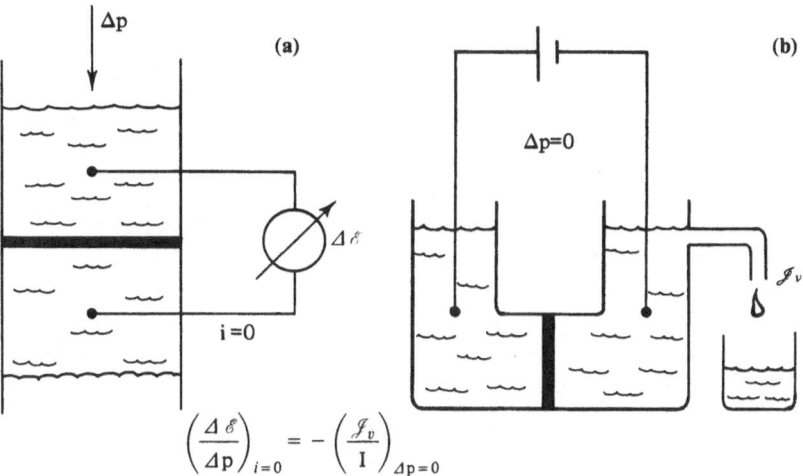

$$\left(\frac{\Delta \mathscr{E}}{\Delta p}\right)_{i=0} = -\left(\frac{\mathscr{I}_v}{I}\right)_{\Delta p=0}$$

Fig. IV-2a, b. Comparison between streaming-potential and electroosmosis measurements (Saxén's law). McKelvey JG Jr, Spiegler KS, Wyllie MRJ (1959) Chem Eng Progr Ser 55: (24) 199

Comparison of Eq. (IV-48) with (IV-49) shows that the streaming potential coefficient is numerically equal to the electroosmotic coefficient (Saxén's law) because of the reciprocity $L_{iv} = L_{vi}$:

$$\left(\frac{\Delta \mathscr{E}}{\Delta p}\right)_{i=0} = -\left(\frac{J_v}{i}\right)_{\Delta p=0}. \tag{IV-50}$$

In similar fashion, we obtain from Eqs. (IV-46) and (IV-47):

$$\left(\frac{\Delta p}{\Delta \mathscr{E}}\right)_{J_v=0} = -\left(\frac{i}{J_v}\right)_{\Delta \mathscr{E}=0}. \tag{IV-51}$$

It is useful to express the flow equations (IV-44) – (IV-47) in terms of readily measurable (or in some cases, tabulated) membrane properties instead of the conductance coefficients, L_{ij} [Eqs. (IV-44), (IV-45)] or resistance coefficients R_{ij} [Eqs. (IV-46), (IV-47)]. This transformation is done in the following.

Expressing $(-\Delta \mathscr{E})$ from Eq. (IV-45), we obtain

$$-\Delta \mathscr{E} = \frac{1}{L_{ii}}(i + L_{iv}\Delta p). \tag{IV-52}$$

We now introduce this parameter into Eq. (IV-44):

$$J_v = \left(L_{vv} - \frac{L_{iv}^2}{L_{ii}}\right)(-\Delta p) + \frac{L_{vi}}{L_{ii}}i. \tag{IV-53}$$

Hence,

$$\left[\frac{J_v}{(-\Delta p)}\right]_{i=0} = \frac{L_{vv}L_{ii} - L_{iv}^2}{L_{ii}} \equiv L_p.$$ (IV-54)

L_p is the *hydraulic permeability* of the membrane, which is always positive.[12] From Eq. (IV-46), we obtain

$$\left(\frac{J_v}{(-\Delta p)}\right)_{i=0} = \frac{1}{R_{vv}}.$$ (IV-55)

Comparison of Eq. (IV-55) with (IV-54) shows that

$$L_p = \frac{1}{R_{vv}} = \frac{L_{vv}L_{ii} - L_{iv}^2}{L_{ii}}.$$ (IV-56)

Introducing the hydraulic permeability, L_p [Eq. (IV-54)] and the electro-osmotic coefficient, β [Eq. (IV-49)] into Eq. (IV-53), we obtain

$$J_v = L_p(-\Delta p) + \beta i.$$ (IV-57)

This *first (transformed) electrokinetic equation* relates the volume flux to the differences of pressure and electric potential, respectively, across the membrane.

Since the *conductance of the membrane per unit surface* (mho cm^{-2}) is defined as

$$\varkappa' = \left(\frac{i}{-\Delta \mathscr{E}}\right)_{\Delta p = 0}$$ (IV-58)

we see from Eq. (IV-45) that

$$\varkappa' = L_{ii}.$$ (IV-59)

Therefore, from Eq. (IV-49):

$$L_{vi} = L_{iv} = \varkappa' \beta.$$ (IV-60)

Substituting Eqs. (IV-59) and (IV-60) in Eq. (IV-45), we obtain the *second (transformed) electrokinetic equation*

---

12 This can be proven from the discussion in Chapter V [see Eqs. (V-3) and (V-4)].

$$i = \varkappa' \beta(-\Delta p) + \varkappa'(-\Delta \mathcal{E}) . \tag{IV-61}$$

A useful transformation of the first flux equation (IV-44) is done in the following manner. First, solve Eq. (IV-54) for L_{vv} and substitute Eqs. (IV-49) and (IV-59); then express L_{vi} by Eq. (IV-60). The result is

$$J_v = (L_p + \varkappa' \beta^2)(-\Delta p) + \varkappa' \beta(-\Delta \mathcal{E}) . \tag{IV-62}$$

This is an alternate form of the *first* (transformed) electrokinetic equation. Comparing it with the *second* (transformed) equation, we see that in this pair of equations, reciprocity is satisfied.[13]

The first transformed equation (IV-57) yields a simple expression, the *electroosmotic counterpressure*. This is the pressure difference which develops across the membrane if, instead of bleeding off the electroosmotic flow, a counterpressure is allowed to build up in the right arm of the U-tube shown in Fig. IV-2(b) until the net volume flow (albeit not the electric current flow) through the diaphragm ceases. In this state, the electroosmotic flow from left to right is counterbalanced by the hydraulic flow from right to left. From Eq. (IV-57), we obtain for the coefficient of electroosmotic counterpressure:

$$\left(\frac{\Delta p}{i} \right)_{J_v = 0} = \frac{\beta}{L_p} . \tag{IV-63}$$

Problems

IV.1. Develop the relationship between the resistance coefficients, R_{ij}, and the mutual friction coefficients, $\zeta_{ij}(i \neq j)$.
Hints: Substitute $J_j = u_j c_j$ in Eqs. (IV-6) and compare the result to the appropriate force balance in the friction model. [Eq. (IV-11) is a typical force balance.]

IV.2. Certain inert wood veneers used as battery separators have been shown to exhibit virtually no transport selectivity toward small anions or cations. In other words, the percentage of the current carried by the cations (or anions) is the same as in free solution.

Using the friction model of flow in porous media, derive flux equations for the ions of a completely dissociated salt composed of two monovalent ions, and of the solvent through a separator of this kind in terms of the forces, concentra-

---

13 As discussed in Chapter V, the generalized driving forces in these equations are properly conjugated to the fluxes; hence, reciprocity is satisfied.

tions and friction coefficients. Verify that reciprocity between the coupling coefficients, L_{ij}, is obeyed, and write down the dimension and units of these coupling coefficients.

IV.3. An ionic membrane is mounted in a tube made from electrically insulating material which also contains two identical reversible electrodes mounted close to the membrane faces. The whole assembly is filled with a dilute electrolyte solution. The following measurements were made:
a) Pressure of 100 atm was applied to the solution on one side of the membrane while the pressure on the other side was atmospheric. The observed flux through the membrane was 2.3×10^{-3} cm s^{-1}, when the electrodes were shorted.
b) The electrical conductance of the membrane was measured. Result (reduced to unit area): $8 \, \Omega^{-1} \, \text{cm}^{-2}$.
c) A potential difference of 8.0 V was applied between the electrodes. This caused an electroosmotic flux (at zero pressure difference) of 3.0×10^{-3} cm s^{-1}.

Write the electrokinetic phenomenological equations for this membrane and calculate the phenomenological coefficients. What streaming-current density do you expect under conditions (a)? The streaming current is the electrical current produced in a wire connecting two reversible electrodes placed close to the low-pressure and high-pressure faces of the membrane, respectively.

Suggestion: It is convenient to operate with electrical units V, A, Ω and W, in addition to cm and s.

1 atm $= 0.1013$ W s cm^{-3}.

Selected Literature

Fundamentals of Flow Coupling

Onsager L (1931) Phys Rev 37:405, ibid 38:2265
Miller DG (1960) Chem Rev 60:15
Haase R (1969) Thermodynamics of irreversible processes. Addison-Wesley, Reading Mass
    The mutual conversion of coordinate frames for flows is one of the many subjects discussed in the latter two references.

Range of Validity of the Linear Flow Equations (IV-5), (IV-6)

Schlögl R, Wiedner G, Woermann D (1975) Transport properties of an ion-exchange membrane far from equilibrium. Ber Bunsenges 79:878

Berg A, Brun TS, Schmitt A, Spiegler KS (1983) Precise transport measurements in membrane systems. In Proc NATO Adv Study Instit Mass transfer and kinetics in ion-exchange. (Maratea Italy 1982) M Nijhoff Publishers, The Hague

Background Literature on Ionic Membranes

Helfferich F (1962) Ion exchange. McGraw-Hill, New York
   This monograph discusses the physical chemistry of ion-exchange materials. *Donnan equilibria* are discussed in detail.
Sélégny E (ed) (1976) Charged gels and membranes I. D Reidel Publishing Co, Dordrecht Holland
   The articles by K Sollner and by T Teorell in this volume summarize the history of this field.
Spiegler KS (1953) On the electrochemistry of ion-exchange membranes. J Electrochem Soc 100: 303 C
Moore DH (ed) (1968) Physical techniques in biological research. 2nd edn, Vol 2A, Chap 7, Electrical potential differences, Academic Press, New York
   This chapter reviews the elements of (equilibrium) electrochemistry and applies them to ionic membranes.

Transport Phenomena in Biological Membranes

Cole KS (1972) Membranes, ions and impulses. University of California Press, Berkeley
Tasaki I (1968) Nerve excitation. Charles C Thomas, Springfield Ill

Friction Model

Einstein A (1905) Ann Phys 17(4):549 (in German)
Lamm O (1944) Arkiv Kemi Min Geol 17A: No 9 1943, ibid 18A: No 2, 9, 10, ibid 18B: No 5 (in English)
   These articles apply the friction model to mass transport in *electrolyte solutions.*
Spiegler KS (1958) Transport in ionic membranes. Trans Faraday Soc 54:1408
   This article derives the basic friction equations for transport of *ions and solvent* in *membranes.*
Kaufmann TG, Leonard EF (1968) Am Inst Chem Eng J 14: (No 1) 110
   Measurements of transport of *sugars* in *cellophane membranes* are described and interpreted in terms of the friction model. This is an example of the use of the friction model for *non-ionic* solutes.
Dawson DF, Thain JF, Meares P (1972) In: Eisenman G (ed) Membranes. Vol 1 Marcel Dekker, New York
   The article contains a critical review of the application of the friction model to membranes.

Electroosmosis

Von Helmholtz A (1879) Ann Phys 7(3):337 (in German)
This classical paper laid the groundwork for the theory of electroosmosis and other manifestations of the coupling of ion flow with solvent flow.
Schmid G (1952) Z Elektrochem 56:181 (in German)
This paper discusses the relative merits of models (e. g., Helmholtz's) and phenomenological approaches [e. g., equations of the type of Eqs. (IV-5) of this text] for the description of coupled mass flows in porous media.
Kimizuka H, Kaibara K, Kumamoto E (1978) Diffusional and conductive membrane permeabilities in the cation-exchange membrane/aqueous electrolyte system. J Membr Sci 4:81
Meares P (1981) Coupling of ion and water fluxes in synthetic membranes. J Membr Sci 8:295

Electroosmotic Velocity Ratio

Cameron RG, Lyle IG, Walker JF, Paterson R (1975) Desalination 17:313
Demisch H-U, Pusch W (1976) Ion-exchange capacity of cellulose-acetate membranes. J Electrochem Soc 123:370
Paterson R (1976) Interpretation and prediction of the transport properties of charged membranes. In: Passino R (ed) Pontificiae Academiae Scientiarum Scripta Varia 40:519, Vatican City

Electroosmosis in Non-Aqueous Media and Electroosmotic Pumping

Phillips RW, Mastrangelo SVR (1972) Electroosmotic energy conversion in the glass/n-propanol system. J Appl Phys 43:3870
Rastogi RP, Srivastava ML, Singh SN (1969) J Phys Chem 73:46

Chapter V. Conductance Coefficients and Reciprocity Relations

A. Laws About the Magnitude of Conductance Coefficients

The development of the flux equations for isothermal coupled transport in membranes, (IV-20) and (IV-21), from the friction model illustrated the linear relationship between fluxes, J_i, J_j, etc. of species i, j, etc. and the generalized (Rayleigh-type) driving forces, F_i, F_j, etc. acting *directly* on these species:

$$J_i = \sum_j L_{ij} F_j \quad \text{(isothermal conditions)} . \tag{V-1}$$

Moreover, it was claimed that since the fluxes, J, vanish in the absence of all generalized driving forces, F, any function $J(F_1, F_2 \ldots)$ can be thus developed, in principle, but the absence of higher terms of the power series, LF^2, $LF^3 \ldots$ can be justified only for very small driving forces, F, i.e., near equilibrium. In general, the range of validity of linear laws of the type of Eq. (V-1) is limited; for instance, for most chemical reactions the linear approximation is adequate only very close to equilibrium. For many transport phenomena, however, the range of approximate linearity extends far beyond equilibrium. For instance, it has been known for many decades that the ion flux in electrolyte solutions is proportional to the applied electric voltage over a considerable voltage range, provided proper electrodes and stirring devices are used. Also, self-diffusion of particles follows a strictly linear relationship between flux and force even in solutions and gases which are very far from isotopic equilibrium. Fick's law of diffusion can be shown to be a linear law in the sense of Eq. (V-1) (Sect. III.C.2.). For isothermal flow of electric current in metals, proportionality between current and applied potential also prevails over considerable ranges of voltages, not just near equilibrium (zero voltage). Even in some *non-isothermal* flow processes, e.g., thermoelectric phenomena (Chap. X), proportionality between fluxes and generalized driving forces is known to hold over considerable ranges of forces,[1] quite far from equilibrium:

$$J_i = \sum_j L_{ij} X_j . \tag{V-2}$$

---

1 Non-linear processes, i.e., flows proceeding under conditions in which this proportionality does not prevail, can lead to oscillations which are of great importance in biology, chemical-reaction kinetics and boiling. These oscillating phenomena are treated in the text by Glansdorff and Prigogine 1971 (see literature list for Chap. I).

Equations (V-1) and (V-2) are of practical value when the conductance coefficients are independent of the forces and represent material constants, or at least can be calculated from material constants (e. g., in some cases, from mutual friction coefficients) and the material composition (e.g., the concentrations of migrating species in membranes). It will indeed be assumed that the flow of each species, i, is described by an equation of the type (V-2), in which the conductance coefficients, L, are independent of the generalized driving forces, X.

The following three laws about the conductance coefficients are discussed in this chapter:

1. *The diagonal conductance coefficients, L_{ii}, are positive:*

$$L_{ii} > 0 . \tag{V-3}$$

2. *The absolute values of the coupling conductance coefficients, L_{ij}, are smaller than the geometric mean of the corresponding diagonal coefficients:*

$$|L_{ij}| < \sqrt{L_{ii}L_{jj}} . \tag{V-4}$$

3. The third law has been illustrated in Chapter IV, viz., *The coupling coefficients are reciprocal [Onsager's reciprocity law (1931)]:*

$$L_{ij} = L_{ji} . \tag{V-5}$$

B. Proofs of the Laws

In the following, the first two laws are proven, while a general proof for the third, which was proven for a special case only [Eqs. (IV-20), (IV-21)], is not presented here.

It is assumed that the transport description has been reduced to the *minimal number of independent fluxes and generalized driving forces.* For instance, in the description of electric-current transport in those cation-exchange membranes, which contain only one kind of electric-current carriers, viz., (positive) cations, it is sufficient to use a flux equation of the type of (V-1) either for the *electric-current* density, i, (A cm^{-2}) or of the cation flux, J_+ (mol cm^{-2} s^{-1}), because the two fluxes are not independent; for all such membranes, one can readily be convered into the other by means of Faraday's law:

$$i = \mathscr{F}Z_+J_+ , \tag{V-6}$$

where $\mathscr{F}$ is Faraday's constant (0.965×10^5 C Eq^{-1}) and Z_+ the valence of the cations (eq mol^{-1}). On the other hand, the flux of water cannot be predicted from either i or J_+ alone on first principles, and is therefore an independent flux for the purpose of the discussion in this chapter.

The numerical magnitudes and directions of the fluxes depend on the generalized driving forces, which must also be a priori independent of each other. It is useful to calculate the *total* entropy creation per unit volume, $\dot{s}$, because certain limitations on the possible values of the conductance coefficients are imposed by the thermodynamic requirement that $\dot{s}$ be positive.

The total entropy production is calculated from Eq. (III-14), by substituting flux expressions from Eq. (V-2) into it:

$$0 < \dot{s} = \sum_i J_i X_i = \sum_i \left(\sum_j L_{ij} X_j \right) X_i \tag{V-7}$$
$$= L_{11} X_1^2 + L_{12} X_1 X_2 + \dots + L_{21} X_1 X_2 + L_{22} X_2^2 + \dots .$$

When a driving force is directly applied on particles 1 (for instance, an electric potential gradient acting of the counterions s in Fig. IV-1) and no other driving force, e. g., hydraulic pressure applied directly on particles 2 (w in Fig. IV-1), all generalized driving forces except X_1 vanish. In this case, only the first term on the right side of Eq. (V-7) is different from zero:

$$0 < (\mathrm{s})_{X_2 = 0, X_3 = 0, \dots} = L_{11} X_1^2 . \tag{V-8}$$

Because X_1^2 is positive, irrespective of the sign of X_1, it follows that L_{11} must be positive also. By similar reasoning, L_{22}, and, in fact, all diagonal conductance coefficients, L_{ii}, are positive, as stated before [Eq. (V-3)]. Although the proof was given here for the special case $X_2 = 0$, the conclusion about L_{22} is valid for any pair of driving forces X_1, X_2. Since the conductance coefficients are assumed to be independent of the generalized driving forces, Eq. (V-7) holds irrespective of the magnitude and sign of the driving forces.

The proof of Eq. (V-4), too, is based on the fact that the total entropy creation, $\dot{s}$, is positive. In case all driving forces except X_1 and X_2 are zero, Eq. (V-7) reduces to

$$0 < (\dot{s})_{X_3 = 0, X_4 = 0, \dots} = L_{11} X_1^2 + (L_{12} + L_{21}) X_1 X_2 + L_{22} X_2^2 . \tag{V-9}$$

The expression on the right side of Eq. (V-8) can be positive only if

$$(L_{11} L_{22} - L_{12} L_{21}) > 0 . \tag{V-10}$$

This can be seen by dividing the expression by X_2^2, thus obtaining $L_{11} z^2 + (L_{12} + L_{21}) z + L_{22}$, $(z \equiv X_1 / X_2)$. The algebraic condition for a positive value of this sum (whether z be positive or negative) is $|(L_{12} + L_{21})| < 2\sqrt{L_{11} L_{22}}$. Assuming reciprocity of the coupling coefficients, this is identical with Eq. (V-10). An analogous proof can be given for any pair of species, i, j, by considering situations in which all generalized driving forces, except X_i and X_j vanish.

It is seen from Eq. (V-2) that the dimensions and units of L_{ij} are those of J_i/X_j, which, in turn are those of $J_iJ_j/\dot{s}$, because it follows from Eq. (III-14) that the dimensions and units of X_j are those of $\dot{s}/J_j$. Similarly, the dimensions and units of L_{ji} are the same as those of $J_jJ_i/\dot{s}$. Thus, it follows that if the fluxes and generalized driving forces are properly conjugated, the *dimensions* (and units) of L_{ij} and L_{ji} are the same. *As for the numerical magnitude* of these coupling coefficients, it has been shown in Chapter IV that for isothermal coupled flows, the magnitudes are also equal [L_{sw} in Eq. (IV-23) compared to L_{ws} in Eq. (IV-24)]. This does not represent a general proof of Eq. (V-5), however, because Eq. (V-5) also applies to flux-force relations not covered by the friction model, e. g., heat fluxes J_Q and chemical-reaction flows $\mathscr{I}_{chem}$ (discussed in Chap. VI) and their conjugated generalized driving forces. By postulating that equations of the type of (V-1) are applicable to *microscopic* local fluctuations (e. g., fluctuations of mass and energy density on a molecular scale), Onsager (1931) brilliantly proved the validity of the reciprocity relations (V-5) on the molecular level and generalized them for macroscopic flow processes in the vicinity of the equilibrium state. Although some reciprocity relations were known before in several fields of science and engineering, e. g., thermoelectricity, electrokinetics and interdiffusion of ideal gases across porous media, Onsager's reasoning, which is not repeated in this book, reduces all these reciprocity relations to a common denominator. The proof can be found in Onsager's original work (1931) and in many treatises of non-equilibrium thermodynamics, e. g., Prigogine's (1967) and Katchalsky and Curran's (1965).

C. Degree of Coupling

The ratio $q \equiv L_{ij}/\sqrt{L_{ii}L_{jj}}$ is a quantitative measure of the degree of coupling between the fluxes of particles i and j. Because of Eq. (V-4), the absolute value of this ratio varies between 0 and 1. The relation of the degree of coupling and the efficiency of energy conversion in coupled flow processes has been discussed in greater detail by Osterle (1964) and by Caplan (1965); as well as Kedem and Caplan (1965).

Problems

V.1. A manufacturer of thermoelectric devices claims that he has developed new thermocouples which satisfy the following phenomenological equations:

$$I = 3.2\,(-\Delta\mathscr{E}) + 0.125\left(-\frac{\Delta T}{\bar{T}}\right)$$

$$\mathscr{J}_Q = 0.030\,(-\Delta\mathscr{E}) + 1.6\times 10^{-5}\left(-\frac{\Delta T}{\bar{T}}\right),$$

where I is the current (A), $\mathscr{J}_Q$ the heat flow (cal s^{-1}), ΔT and $\Delta\mathscr{E}$ the small temperature difference (°C) and voltage between the junctions, respectively. T is the absolute temperature.
Examine this claim in the light of the laws about conductance coefficients and state why you believe it is or is not valid.
1 W s = 0.2389 cal.

V.2. Calculate the coupling coefficient for an ionic membrane which has the following flow constants:

Hydraulic permeability (when $\Delta\mathscr{E} = 0$), $L_{vv} = 2.3\times 10^{-4}$ cm s^{-1} MPa^{-1}
Electric conductance per unit area, $L_{ii} = 8\ \Omega$ cm^{-2}
Electroosmotic velocity, $L_{vi} = 3.75\times 10^{-4}$ cm s^{-1} V^{-1}

Hints: The coupled hydraulic[2] and electric fluxes are described by Eqs. (IV-44) to (IV-47).

Selected Literature

Onsager L (1931) Phys Rev 37:405, ibid 38:2265
Prigogine I (1967) See literature list for Chapter I
Katchalsky A, Curran P (1965) See literature list for Chapter I
Osterle JF (1964) A unified treatment of the thermodynamics of steady-state energy conversion. Appl Sci Res A 12
Caplan SR (1965) The degree of coupling and the efficiency of fuel cells and desalination processes. J Phys Chem 69:3801
Kedem O, Caplan SR (1965) Degree of coupling and its relation to efficiency of energy conversion. Trans Faraday Soc 61:1987
Odum HT, Pinkerton RC (1955) Time's speed regulator: the optimum efficiency for maximum power output in physical and biological systems. Am Sci 43:331

---

2 Flux of non-compressible fluid.

Chapter VI. General Energetics of Chemical Reactions

A. Introduction

Previous chapters dealt with the relationship between driving forces and flows of different nature. All of these were *vectorial*. Frequently, complex systems, such as living cells or industrial plants, also encompass zones in which the *scalar* processes of chemical reactions take place. To include these chemical-reaction events in the energetic analysis of the systems, it is necessary to fit chemical reactions into the force-flow scheme and the flow-coupling relationship developed for vectorial flows. This is indeed possible and is discussed in this chapter. First, the chemical analogues to vectorial flows and generalized driving forces are defined, followed by discussion of reciprocity relations in a system of simple, coupled chemical reactions.

B. Generalized Forces and "Flows" of Chemical Reactions

1. Rate of Reaction Advancement

Consider a closed vessel in which a chemical reaction proceeds according to the stoichiometric formula:

$$(-v_A)A + (-v_B)B + \ldots \rightarrow v_C C + v_D D + \ldots$$

$$(v_A, v_B, \ldots < 0; \; v_C, v_D, \ldots > 0).$$

(VI-1)

The *stoichiometric coefficients*, v_i, are taken positive for reaction products, and negative for reactants. For instance, for the ammonia synthesis $N_2 + 3H_2 \rightarrow 2NH_3$, $v_{N_2} = -1$, $v_{H_2} = -3$, $v_{NH_3} = +2$.

The vessel is so large that the change of temperature, T, and pressure, p, and the chemical potentials of all components due to the reaction of v_A, $v_B \ldots$ mol of components A, B $\ldots$ is negligible. The *exergy*, Λ, of the contents of this vessel, with respect to a reservoir at temperature $T_0 = T$, and pressure, $p_0 = p$ is equal to the *Gibbs free energy*, G, of its contents:

$$\Lambda = G = \mu_A^c n_A(t) + \mu_B^c n_B(t) + \ldots + \mu_C^c n_C(t) + \mu_D^c n_D(t) + \ldots . \tag{VI-2}$$

n_i is the number of mols of component i. Contrary to the convention about the stoichiometrical coefficients, ν_i, n_i is taken positive for both reaction products and reactants.

Since the chemical potential, μ_i, is the partial molal free energy of component i, the free energy of the reacting mixture at time t, which is equal to its exergy with respect to the specific reservoir chosen here, is:

$$\Lambda = \sum_i \mu_i^c n_i(t) . \tag{VI-3}$$

The rate of exergy disappearance due to this isothermal chemical reaction is:

$$\dot{\Lambda} = \frac{d\Lambda}{dt} = \mu_A^c \frac{dn_A}{dt} + \mu_B^c \frac{dn_B}{dt} + \ldots + \mu_C^c \frac{dn_C}{dt}$$

$$+ \mu_D^c \frac{dn_D}{dt} + \ldots = \sum_i \mu_i^c \frac{dn_i(t)}{dt} < 0 . \tag{VI-4}$$

In principle, the rate of the reaction can be defined as the rate of creation of any reaction product, e. g., dn_B/dt, or the rate of consumption of any reactant, e. g., dn_A/dt. In order to unify these different parameters, however, de Donder defined a *rate of reaction advancement* $d\xi/dt$:

$$\frac{d\xi}{dt} \equiv \frac{1}{\nu_C} \frac{dn_C}{dt} = \frac{1}{\nu_D} \frac{dn_D}{dt} = \ldots = \frac{1}{\nu_A} \frac{dn_A}{dt}$$

$$= \frac{1}{\nu_B} \frac{dn_B}{dt} = \ldots = \frac{1}{\nu_i} \frac{dn_i}{dt} > 0 . \tag{VI-5}$$

The rate of reaction advancement is always positive since, for the reaction products, ν_i is always positive and dn_i/dt is positive also as the reaction proceeds, while for the reactants, both ν_i and dn_i/dt are negative.

2. Affinity

By substituting for the dn_i/dt in Eq. (VI-4) from Eq. (VI-5), one can express the exergy disappearance in the isothermal chemical reaction in terms of the rate of reaction advancement:

$$\dot{\Lambda} = (\mu_A^c \nu_A + \mu_B^c \nu_B + \ldots + \mu_C^c \nu_C + \mu_D^c \nu_D + \ldots) \frac{d\xi}{dt} = \left(\sum_i \nu_i \mu_i^c \right) \frac{d\xi}{dt} < 0 . \tag{VI-6}$$

To obtain the generalized driving force for the chemical reaction, the rate of entropy creation is calculated from the Maxwell-Gouy-Stodola equation (II-33). Considering that in this case the reservoir temperature, T_0, is equal to the temperature of the reacting mixture, we obtain from Eq. (VI-6):

$$- \dot{A} = T\dot{S}_{\text{chem}} = - \left(\sum_i \nu_i \mu_i^c \right) \frac{d\xi}{dt}. \tag{VI-7}$$

Therefore, the rate of entropy creation is

$$\dot{S}_{\text{chem}} = \frac{d\xi}{dt} \sum_i \left(- \frac{\nu_i \mu_i^c}{T} \right). \tag{VI-8}$$

It is seen that the rate of entropy creation is the product of two terms. If we consider the rate of reaction advancement as the flow of the chemical reaction,

$$\mathscr{I}_{\text{chem}} \equiv \frac{d\xi}{dt}, \tag{VI-9}$$

then the generalized driving force is [Eq. (III-1)]

$$_\Delta X_{\text{chem}} \equiv \dot{S}_{\text{chem}} / \mathscr{I}_{\text{chem}} = - \frac{1}{T} \sum_i \nu_i \mu_i^c. \tag{VI-10}$$

Because isothermal reactions are considered here, T is constant; it is useful to define instead a Rayleigh-type driving force, $F \equiv XT$, as was done for isothermal vectorial flow phenomena (Table III-1). De Donder named this generalized driving force for isothermal chemical reactions *affinity*, $\mathscr{A}$:

$$\mathscr{A} \equiv T_\Delta X_{\text{chem}} = - \sum_i \nu_i \mu_i^c. \tag{VI-11}$$

For instance, for the ammonia synthesis:

$$- \sum_i \nu_i \mu_i^c = \mu_{N_2}^c + 3\mu_{H_2}^c - 2\mu_{NH_3}^c. \tag{VI-12}$$

When the reaction advances from left to right, the rate of reaction advancement [Eq. (VI-9)] is positive. If the reaction is to advance *spontaneously*, exergy disappears ($\dot{A} < 0$). Substituting $\mathscr{A}$ [Eq. (VI-11)] into Eq. (VI-6), it is seen that for spontaneous reactions, the affinity is positive.

When the reaction mixture reaches equilibrium, the affinity is zero:

$$\left(\sum_i v_i \mu_i^c \right)_{eq} = - \mathscr{A}_{eq} = 0 . \tag{VI-13}$$

For ideal gases, this *equilibrium condition* can readily be expressed in terms of the concentrations of the components, i, by substituting for the chemical potentials from Eq. (III-8):

$$\left(\sum_i v_i RT \ln c_i \right)_{eq} = - \sum_i v_i \mu_i^0 . \tag{VI-14}$$

Hence

$$(c_1^{v_1} \cdot c_2^{v_2} \cdot c_3^{v_3} \ldots)_{eq} = \exp \left(\frac{- \sum_i v_i \mu_i^0}{RT} \right) \equiv (K_{eq})_T . \tag{VI-15}$$

The expression on the right is a characteristic *equilibrium constant* for any chemical reaction at a given temperature because the standard chemical potentials, μ_i, are constants. For instance, for the ammonia synthesis reaction in the ideal-gas range, Eq. (VI-15) is[1]

$$\left(\frac{c_C^{|v_C|}}{c_A^{|v_A|} \ c_B^{|v_B|}} \right)_{eq} = \left(\frac{c_{NH_3}^2}{c_{N_2} c_{H_2}^3} \right)_{eq} = (K_{eq})_{(N_2 + 3H_2 \rightleftharpoons 2NH_3), T} . \tag{VI-16}$$

This is one of the possible descriptions of the equilibrium for chemical reactions of ideal gases.

To express affinity in terms of the concentrations, rather than the chemical potentials, we substitute for the latter from Eq. (III-8):

$$\mathscr{A} = - \sum_i v_i \mu_i^c = - \left(\sum_i v_i \mu_i^0 + \sum_i RT \ln c_i^{v_i} \right) . \tag{VI-17}$$

We can express the algebraic sum of the standard chemical potentials in terms of the equilibrium constant of the reaction by means of Eq. (VI-15):

$$\sum_i v_i \mu_i^0 = - RT \ln K_{eq} . \tag{VI-18}$$

---

[1] In industrial practice, the synthesis is performed at high pressures; since the ideal-gas laws are not valid under these conditions, activities have to be used instead of concentrations. Since the activity coefficients depend to some extent on the total pressure, the equilibrium constant, too, depends on the total pressure in this case.

Expressing the equilibrium constant in terms of the concentrations of reaction products and reactants [also by Eq. (VI-15)], we obtain an expression for the affinity of ideal-gas reactions in terms of these concentrations:

$$\mathscr{A} = - \left(\sum_i (RT \ln c_i^{\nu_i}) - RT \ln K_{eq} \right) = - RT \sum_i \nu_i \ln \left(\frac{c_i}{c_{i,eq}} \right), \qquad \text{(VI-19)}$$

where $(c_i)_{eq}$ is the final concentration of component i, when equilibrium is established.

3. Affinity and Reaction Rate Close to Equilibrium

For any component, i, present at concentration c_i in a mixture reacting isothermally, the *deviation from equilibrium* may be characterized by

$$c_i - (c_i)_{eq} \equiv {}_{eq}\Delta c_i. \qquad \text{(VI-20)}$$

When this deviation is small compared to c_i, we use the approximation (first term of the Taylor expansion of the ln-function):

$$\ln \left(1 + \frac{{}_{eq}\Delta c_i}{c_{i,eq}} \right) \simeq \frac{{}_{eq}\Delta c_i}{c_{i,eq}}. \qquad \text{(VI-21)}$$

By introducing this approximation into Eq. (V-19), we obtain

$$\mathscr{A} \simeq - RT \sum_i \nu_i \ln \left(1 + \frac{{}_{eq}\Delta c_i}{c_{i,eq}} \right) \simeq - RT \sum_i \frac{\nu_i}{c_{i,eq}} {}_{eq}\Delta c_i. \qquad \text{(VI-22)}$$

It is seen that near equilibrium, the affinity is an algebraic sum of terms, each of which is proportional to the respective deviation from equilibrium.

4. Isomerization Reactions Near Equilibrium

Because of the complexity of many chemical reactions, which often involve one or several intermediate products in the sequence of events leading from reactants to reaction products, the rate of advancement of an overall reaction (VI-1) can, in general, not be expected to be simply proportional to the affinity, when both are calculated from the stoichiometric reaction formula (VI-1) by use of Eqs. (VI-5) and (VI-11), respectively. To illustrate the existence of reciprocity relationships in the rate equations for coupled reactions (Sect. VI.C), we examine the phenomenology of isomerization (molecular rearrangement) reactions; the stoichiometric formulae for such reactions are particularly simple,[2] being represented by reaction equations of the type:

$$A \rightarrow B. \tag{VI-23}$$

The affinity for this reaction is calculated from Eq. (VI-11):

$$\mathscr{A}_{A \rightarrow B} = \mu_A^c - \mu_B^c. \tag{VI-24}[3]$$

At equilibrium, the generalized driving force for the reaction vanishes:

$$0 = (\mu_B^c)_{eq} - (\mu_A^c)_{eq}. \tag{VI-25}$$

Adding Eq. (VI-25) to (VI-24), we obtain:

$$\mathscr{A}_{A \rightarrow B} = [\mu_A^c - (\mu_A^c)_{eq}] - [\mu_B^c - (\mu_B^c)_{eq}] = {}_{eq}\Delta\mu_A^c - {}_{eq}\Delta\mu_B^c, \tag{VI-26}$$

where

$$_{eq}\Delta\mu^c \equiv \mu^c - \mu_{eq}^c. \tag{VI-27}$$

Near equilibrium, the affinity may be expressed by Eq. (VI-22):

$$\mathscr{A}_{A \rightarrow B} \simeq \left(\frac{RT}{c_{A,eq}}\right){}_{eq}\Delta c_A - \left(\frac{RT}{c_{B,eq}}\right){}_{eq}\Delta c_B \tag{VI-28}$$

and it is also assumed that the rate of progression of the reaction is roughly proportional to the affinity:[4]

$$\mathscr{J}_{chem} \simeq \alpha_1 \mathscr{A} \quad \text{(near equilibrium)}, \tag{VI-29}$$

where α_1 is the factor of proportion.

---

2 The simplicity of the stoichiometric formula for an isomerization does not guarantee that there are no intermediate steps, nor that the rate equations are simple. We shall assume, however, [Eq. (VI-29)] that within the range of reaction conditions considered here, the rate of advancement of the isomerization is proportional to its affinity.
3 Note the analogy with isothermal diffusion (Table III-1). In that process, too, the generalized driving force is a chemical-potential difference.
4 The range of validity of this assumption varies with the nature of the isomerization reaction, and is usually narrow.

Hence, Eq. (VI-26) yields:

$$\mathscr{I}_{chem} = \alpha_1 (_{eq}\Delta\mu_A^c - {}_{eq}\Delta\mu_B^c) \quad \text{(near equilibrium)} \tag{VI-30}$$

or, in terms of concentrations [Eq. (VI-28)]

$$\mathscr{I}_{chem} = \left(\frac{RT\,\alpha_1}{c_{A,eq}}\right)_{eq}\Delta c_A - \left(\frac{RT\,\alpha_1}{c_{B,eq}}\right)_{eq}\Delta c_B \quad \text{(near equilibrium)}. \tag{VI-31}$$

Thus, at least in the vicinity of the isotopic equilibrium, the rate of isomerization is linearly related to the deviations of the isomer concentrations from equilibrium.

5. Comparison of Isomerization-Rate Equation to Conventional Chemical Kinetics

This section demonstrates that the rate equation (VI-31), derived in the previous sections of this chapter from consideration of the rate of exergy disappearance [Eq. (VI-7)] as a product of rate of reaction advancement and affinity, is compatible with conventional chemical kinetics of (isothermal) first-order reactions, i. e., the rates of isomerization, $j_{A\to B}$ and $j_{B\to A}$, are proportional to the concentrations, c_A and c_B, respectively:[5]

$$j_{A\to B} = k_{A\to B}c_A \tag{VI-32}$$

and

$$j_{B\to A} = k_{B\to A}c_B. \tag{VI-33}$$

The *net* rate of formation of B is:

$$j_B \equiv (dc_B/dt) = j_{A\to B} - j_{B\to A} = k_{A\to B}c_A - k_{B\to A}c_B. \tag{VI-34}$$

At equilibrium, the rates of forward and backward reaction are equal.

$$(j_{A\to B})_{eq} = (j_{B\to A})_{eq} \tag{VI-35}$$

or, in terms of Eqs. (VI-32) and (VI-33):

$$k_{B\to A}c_{B,eq} - k_{A\to B}c_{A,eq} = 0. \tag{VI-36}$$

---

5 j represents the rate of isomer production in unit volume of the reaction mixture (mol s^{-1} cm^{-3}).

Add Eqs. (VI-34) and (VI-36):

$$j_B = k_{A \to B}(c_A - c_{A,eq}) - k_{B \to A}(c_B - c_{B,eq}) = (k_{A \to B})_{eq}\Delta c_A - (k_{B \to A})_{eq}\Delta c_B.$$
$$\text{(VI-37)}$$

Comparing Eq. (VI-37), derived from elementary chemical kinetics, to Eq. (VI-31), it is seen that the two equations are indeed of the same form, relating the deviations of the isomer concentrations from their equilibrium concentrations to the net rate of isomerization by means of rate constants, $k_{X \to Y} = RT\alpha_1/c_{X,eq}$.

C. Coupled Chemical Reactions

1. Conjugation of Generalized Driving Forces with Rates of Reactions

In a mixture of components undergoing chemical reactions, the rates of certain reactions may depend on the progress of others, i.e., there may be *coupling* between chemical reactions. It is important to determine whether reciprocity relations similar to Eq. (IV-19) derived for coupled transport processes exist, because the validity of such relations reduces the number of parameters to characterize the system from a chemical-kinetic viewpoint, and consequently also decreases the labor for obtaining the experimental data necessary for this purpose. In this section, it is shown that reciprocity relations between kinetic coefficients do indeed exist near equilibrium. These coefficients are again functions of rate constants and concentrations, as in Eq. (IV-20). The simple model of reaction coupling is the *isomerization triangle* (Onsager 1931). This is a mixture of three isomers, A, B, and C, respectively, which can convert into each other through molecular rearrangement (Fig. VI-1) by different reaction pathways. For the sake of this demonstration, we assume that the mixture exhibits ideal-gas behavior, and is held in a container of unit volume placed in a thermostat at temperature

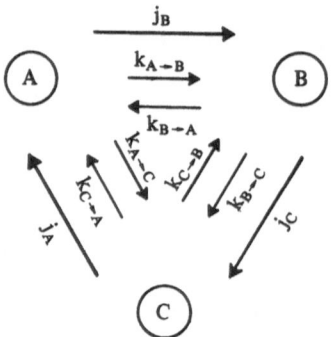

Fig. VI-1. Isomerization triangle. *A* is assumed to isomerize directly to *B*, or first change into *C*, which, in turn, isomerizes to *A*. Similarly, *B* or *C* can change into another isomer either directly or by a two-step process. The temperature, $T = T_0$ is uniform and constant

T_0, so that rate equations developed in Sect. VI.B.4 and VI.B.5 may be applied here. In fact, the reciprocity relations to be demonstrated are valid not only for isomerizations, but for other coupled reactions proceeding by different pathways (Wegscheider 1901, Skrabal 1950).

The affinities for the isomerization reactions $A \rightarrow B$, $B \rightarrow C$, and $C \rightarrow A$, respectively, are expressed in terms of the chemical potentials of the components by Eq. (VI-24):

$$\mathscr{A}_{A \rightarrow B} = \mu_A^c - \mu_B^c \tag{VI-38}$$

$$\mathscr{A}_{B \rightarrow C} = \mu_B^c - \mu_C^c \tag{VI-39}$$

$$\mathscr{A}_{C \rightarrow A} = \mu_C^c - \mu_A^c . \tag{VI-40}$$

The net reaction of the $A \rightarrow B$ isomerization, j_B, is defined as the rate of creation of B from A minus the rate of decomposition of B to A. j_C (for $B \rightarrow C$) and j_A (for $C \rightarrow A$) are similarly defined in accordance with the isomerization scheme shown in Fig. VI-1. Because of the reaction coupling shown in this figure, neither the rates of the three reactions nor the driving forces are completely independent of each other. In fact, it follows from Eqs. (VI-38) to (VI-40) that

$$\mathscr{A}_{C \rightarrow A} = - (\mathscr{A}_{A \rightarrow B} + \mathscr{A}_{B \rightarrow C}) . \tag{VI-41}$$

Thus, the number of independent, generalized forces can be reduced to two, and so can the number of independent conjugated flows. To determine these flows, we calculate first the total entropy production per unit time and unit volume:

$$\dot{s} = \sum_k \dot{s}_k = \frac{1}{T} \sum_k F_k j_k = \frac{1}{T} (j_B \, \mathscr{A}_{A \rightarrow B} + j_C \, \mathscr{A}_{B \rightarrow C} + j_A \, \mathscr{A}_{C \rightarrow A}) . \tag{VI-42}$$

Multiply this equation by the temperature, T, and substitute for the affinity, $\mathscr{A}_{C \rightarrow A}$, from Eq. (VI-41):

$$T\dot{s} = (j_B - j_A) \, \mathscr{A}_{A \rightarrow B} + (j_C - j_A) \, \mathscr{A}_{B \rightarrow C} . \tag{VI-43}$$

Generalizing the formalism discussed in Chapter III, in which $T\dot{s}$ for certain single-flow processes was found to be the product of the flux with its conjugated driving force [Eq. (III-15)], and considering that in the presence of several irreversible processes, the entropy productions of each single process add up to the *total* entropy production in the system, as was done in Sect. (V.B), we consider the sum of the two products on the right side of Eq. (VI-43) as conjugated flux-force pairs. Since the reactions are coupled, it is also assumed that linear flow equations of the type (V-1) are valid, at least near equilibrium:

$$j_B - j_A = L_{11}\,\mathcal{A}_{A\to B} + L_{12}\,\mathcal{A}_{B\to C} = -\frac{dc_A}{dt} \qquad\qquad (VI\text{-}44)$$

$$j_C - j_A = L_{21}\,\mathcal{A}_{A\to B} + L_{22}\,\mathcal{A}_{B\to C} = \frac{dc_C}{dt}. \qquad\qquad (VI\text{-}45)$$

Thus, the description of *three* coupled equations can be reduced to rate equations involving only *two* independent, generalized driving forces and flow parameters.

Because of the thermodynamic considerations discussed in Sect. V.B, the conductance coefficients, L, obey the basic laws (V-3) and (V-4). As for the reciprocity relation, (V-5), it will now be shown that it is valid also.

2. Reciprocity in Coupled Isomerization Reactions

In principle, one can imagine two ways to achieve constant concentrations in a mixture of isomers A, B, and C at constant and uniform temperature and pressure, namely (a) three separate equilibria (Fig. VI-2a) or (b) a steady state (round-the-cycle scheme), in which each species is created as rapidly by one reaction as it decomposes by the subsequent one. For instance, B is created by the A → B reaction at the same rate as it decomposes by the B → C reaction (Fig. VI-2b).

Chemists have traditionally assumed that in the state of genuine equilibrium, three equilibria exist, rather than a round-the-cycle steady state.[6] Onsager, Tolman and others have shown that the existence of separate equilibria, also called the principle of detailed balance, is a consequence of the statistical-mechanical principle of microscopic reversibility.

---

6 The principle of separate equilibria can be made plausible by reasoning that the other alternative (round-the-cycle movement) could lead to an impossible conclusion. At equilibrium, two reversible electrodes, specific to (i. e., whose electric potential depends on) the concentration of B and of C, respectively, are introduced. (Specific electrodes, having electric potentials determined by one species only of a mixture, and not by others, are well-known in chemistry; there is no fundamental reason against their existence for isomers A and B, respectively.) Because the isothermal, isobaric reaction mixture A, B, C is at equilibrium, it can produce no work. Hence there is no voltage between the electrodes.

   If there are separate equilibria, the addition of a specific catalyst for the B→C reaction cannot change the equilibrium concentration $(c_B)_{eq}$, since catalysts cannot affect the percentage of B and C at equilibrium. If a steady state caused by a round-the-

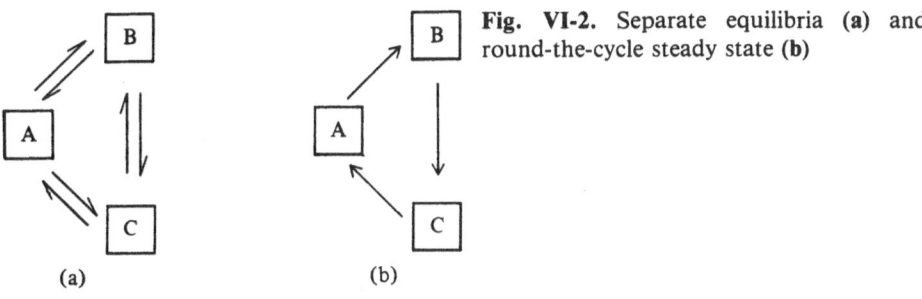

Fig. VI-2. Separate equilibria **(a)** and round-the-cycle steady state **(b)**

(a) (b)

Because each of the three isomerization reactions is at equilibrium, forward rates are equal to backward rates in all three.

$$(j_{A \to B})_{eq} = (j_{B \to A})_{eq}; \quad (j_{B \to C})_{eq} = (j_{C \to B})_{eq}; \quad (j_{C \to A})_{eq} = (j_{A \to C})_{eq}. \quad \text{(VI-46)}$$

Assuming validity of rate equations of the type (VI-36), we obtain *for the equilibrium state:*

$$k_{A \to B} c_{A, eq} = k_{B \to A} c_{B, eq} \qquad \text{(VI-47)}$$

$$k_{B \to C} c_{B, eq} = k_{C \to B} c_{C, eq} \qquad \text{(VI-48)}$$

$$k_{C \to A} c_{C, eq} = k_{A \to C} c_{A, eq}. \qquad \text{(VI-49)}$$

In an isomer mixture *in the vicinity of the equilibrium state*, the affinities of the reactions do not vanish. The affinity of the $A \to B$ reaction is expressed by Eq. (VI-28). On substituting in it for $c_{B, eq}$ from Eq. (VI-47), we obtain

$$\mathscr{A}_{A \to B} = \frac{RT}{k_{A \to B} c_{A, eq}} \underbrace{[k_{A \to B}(_{eq}\Delta c_A) - k_{B \to A}(_{eq}\Delta c_B)]}_{j_B}. \qquad \text{(VI-50)}$$

Comparing Eq. (VI-50) to (VI-37), it is seen that the expression in parentheses is the net rate of formation of B, defined in Eq. (VI-34). Solving (VI-50) for j_B, we obtain

---

cycle movement kept the concentrations of B constant before the catalyst is added, however, the addition of the catalyst will enhance the rate of decomposition of B to C and thus change $(c_B)_{eq}$, at least temporarily. The B-specific electrode senses this change, responds by change of its electric potential, and a voltage appears between the two electrodes. If the system were originally at true thermodynamic equilibrium, it would not be possible to cause this effect by unchanging external devices (Postulate 1, Chap. I). Therefore, the round-the-cycle scheme (Fig. VI-2b) does not represent thermodynamic equilibrium.

$$j_B = \frac{k_{A \to B} c_{A,eq}}{RT} \mathscr{A}_{A \to B} \cdot \qquad\qquad\qquad (VI\text{-}51)$$

In analogy, the net rates for formation of C and A are calculated from Eqs. (VI-48) and (VI-49), respectively:

$$j_C = \frac{k_{B \to C} c_{B,eq}}{RT} \mathscr{A}_{B \to C} \qquad\qquad\qquad (VI\text{-}52)$$

$$j_A = \frac{k_{C \to A} c_{C,eq}}{RT} \mathscr{A}_{C \to A} = - \frac{k_{C \to A} c_{C,eq}}{RT} (\mathscr{A}_{A \to B} + \mathscr{A}_{B \to C}). \qquad (VI\text{-}53)$$

To test for reciprocity in the rate equations (VI-44), (VI-45), we substitute in them the net fluxes from Eqs. (VI-51) to (VI-53), which we derived from the principle of separate equilibria:

$$j_B - j_A = \frac{k_{A \to B} c_{A,eq} + k_{C \to A} c_{C,eq}}{RT} \mathscr{A}_{A \to B} + \frac{k_{C \to A} c_{C,eq}}{RT} \mathscr{A}_{B \to C} \qquad (VI\text{-}54)$$

$$j_C - j_A = \frac{k_{C \to A} c_{C,eq}}{RT} \mathscr{A}_{A \to B} + \frac{k_{B \to C} c_{B,eq} + k_{C \to A} c_{C,eq}}{RT} \mathscr{A}_{B \to C}. \qquad (VI\text{-}55)$$

It is seen that in this system of coupled isomerization reactions, the reciprocity relations hold indeed in the vicinity of the equilibrium state, where the approximations introduced in Sects. VI.B.3 and VI.B.4 are valid. While this finding imposes limitations on the range of validity of these useful relations, it should be noted that this reciprocity may be expected in the rate equations, not only of isomerization reactions, but also of other reactions which can proceed along different paths. Because of the complexity of most such reactions, and the resulting need for well-chosen chemical-kinetic data for their complete kinetic interpretation, few complete sets of data for systematic characterization of such reaction groups along the lines discussed here are available. Two sets of data for an isomerization triangle, uncatalyzed and uncatalyzed, respectively, are included in Problem VI.3.

Problems

VI.1. Calculate the heat of combustion of carbon monoxide (CO) at 1000°C, 1 atm, from the following literature data:

At 18 °C, 1 atm: $2CO(g) + O_2(g) \rightarrow 2CO_2(g) + 135220$ cal
Molar specific heats, $\tilde{C}_p$ (cal mol^{-1}):
For O_2 and CO: $\tilde{C}_p = 6.50 + 0.001\ T$
For CO_2: $\tilde{C}_p = 7.00 + 0.0071\ T - 0.00000186\ T^2$

VI.2. At 25 °C the vapor pressure of pure iodine is 0.305 mm, its solubility in water is 0.00132 mol 1^{-1} and its distribution ratio between carbon tetrachloride and water is 86.
Write the functional relationship (at 25 °C) between:
a)   Molar free energy of iodine vapor as a function of its pressure
b, c) Molar free energy of dissolved iodine as a function of its concentration,
      b) in water and c) in carbon tetrachloride.
The *numerical constants* should be chosen such that free energies are in W s mol^{-1}, vapor pressure in atmosphere, concentrations in mol 1^{-1}.
Hint: Take solid iodine at 25 °C as the standard state. Assume ideal-gas law for the vapor and ideal-solution law for the solutions.

VI.3. The rate of isomerization of the three butenes C_3H_8 has been studied [Haag WO, Pines H (1960) J Am Chem Soc 82:387, ibid p 2488].
The following rate constants were found by measuring isomerization rates near equilibrium:

	With catalyst Al_2O_3	With a mixed catalyst $Na - Al_2O_3$
k_{12}/k_{13}	2.4	4.0
k_{23}/k_{21}	1.0	0.769
k_{31}/k_{32}	0.4	0.277

Test which, if any, of these two groups of experimental results agree with the statement that near equilibrium each of the three reactions is separately in equilibrium.

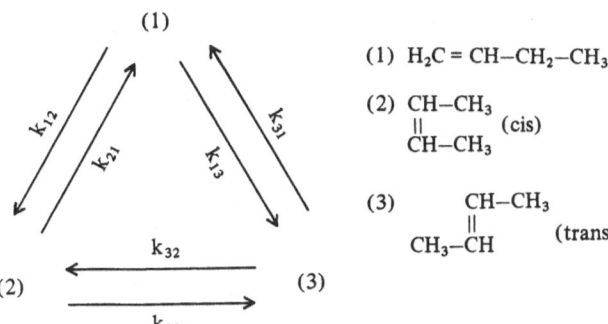

(1) $H_2C = CH-CH_2-CH_3$

(2) CH–CH$_3$
    ‖   (cis)
    CH–CH$_3$

(3)   CH–CH$_3$
          ‖   (trans)
    CH$_3$–CH

Fig. VI-3. Isomerization of butenes

Selected Literature

Phenomenology of Chemical Reactions

De Donder Th, Van Rysselberghe P (1936) Affinity. Stanford University Press, Menlo
  Park California
Denbigh K (1951) The thermodynamics of the steady state. Methuen, London

Principle of Separate Equilibria (Detailed Balance)

Wegscheider R (1901) Monatsh Chem 22:849
Skrabal A (1950) Monatsh Chem 81:239

Principle of Microscopic Reversibility

Onsager L (1931) Phys Rev 37:405, ibid 38:2265
Tolman RC (1938) The principles of statistical mechanics. Oxford University Press, New
  York

Isomerization Triangle

Maurel R, Miller DG (1962) C R Acad Sci (Paris) 255:1266

Chapter VII. Interdiffusion of Gases in Porous Media

A. Introduction

Many industrial processes involve interdiffusion of gases. Gas separation by selective diffusion through porous media and membranes is based on the selective permeability of these media to different molecular species. Moreover, gas diffusion to and from reaction sites in porous catalysts, porous fuel-cell electrodes or other condensed phases is sometimes the rate-controlling step in mass-transfer processes.

While advanced theories for diffusion in free space exist, which make it possible to predict diffusion coefficients and their dependence on different variables (e. g., temperature and pressure), the prediction of diffusion rates in porous media and membranes is stymied by the difficulty of introducing the complex geometry of the porous medium into the equations of the kinetic theory of gases. Attempts have been made to represent specific models of porous media, e. g., the dusty-gas model, in which the porous medium is represented as an additional gas, the molecules of which are stationary, or the capillary-bundle model. Since the geometry of real porous media is usually much more complicated, however, it is worthwhile to examine the potential of a *phenomenological* approach, which can at least describe the dependence of diffusivities (diffusion coefficients) on such parameters as temperature and pressure, although it cannot make a priori predictions of the magnitude of diffusivities. In this chapter, it is shown that many useful relationships in the field of gas interdiffusion can indeed be obtained by the elementary methods described in Chapter IV.

We limit all subsequent considerations to ideal gases at uniform temperature. Absorption and chemical interactions are neglected. Moreover, only diffusion is considered here. Permeation under applied pressure gradients is not treated. Despite all these limitations, a variety of macroscopic effects can be observed:[1]

When two gases, A and B, held in two separate and closed vessels and originally at equal pressure, interdiffuse across a porous diaphragm, the transport rate depends on the characteristic pore dimension (pore size). If the pore size is very large compared to the free path (normal diffusion range), the gases diffuse

---

1 Summarized by the author in Ind. Eng. Chem. 5:529 (1966); the treatment in this chapter follows the development in this paper.

at almost equal rates in opposite directions. In other words, the fluxes are completely coupled and characterized by a single interdiffusion coefficient, D_{AB}, and only a very small (if any) pressure gradient develops between the vessels. On the other hand, if the pore size is smaller than the free path (Knudsen diffusion range), each gas diffuses at the beginning at its own independent rate, characterized by individual diffusion coefficients, D_{AK} and D_{BK}, respectively, and an appreciable transient pressure gradient develops spontaneously. Different diffusion rates are also observed in the steady-state interdiffusion of gases across a porous plug, whose terminal faces are flushed with the two gases at equal pressure. In self-diffusion[2] experiments on gas A the equal diffusion coefficients, $\bar{D}_A^a$, of the interdiffusing species are given by

$$\frac{1}{\bar{D}_A^a} = \frac{1}{\bar{D}_{A,K}} + \frac{1}{\bar{D}_{AA^*}}, \tag{VII-1}$$

where $\bar{D}_{A,K}$ is the Knudsen diffusion coefficient of A through the porous medium which is independent of the pressure (as long as the free path is made larger than the pore size), and $\bar{D}_{AA^*}$ is a coefficient related to the self-diffusion coefficient of gas A in free space and inversely proportional to the pressure. The physical model corresponding to Eq. (VII-1) is that of two diffusion resistances in series; a plot of $1/\bar{D}_A^a$ vs. pressure is linear with a positive intercept. This equation is known as *Bosanquet's equation.*

It is pertinent to ask what minimal assumptions are necessary to explain these phenomena and to derive the basic laws of gas interdiffusion across porous media, such as Eq. (VII-1), or the fact that the steady-state interdiffusion flux ratio for binary mixtures (at absolutely uniform pressure) is always equal to the ratio of the Knudsen diffusion coefficients: This is true at all pressures, not just in the Knudsen region.

This chapter is to aid in the *phenomenological*, rather than the molecular, understanding of interdiffusion phenomena, both in gases and solutions. For *molecular* theory, the reader is referred to texts on the kinetic theory of gases.

B. Friction Model

When two ideal gases, A and B, interdiffuse across a porous diaphragm *at uniform total pressure* and temperature, the generalized driving force acting on one mol of A is the gradient of the chemical potential of A (Table III-1):

---

2 Self-diffusion is the interdiffusion of very similar species, which have the same, or almost the same, diffusivity, e. g., ortho vs parahydrogen, or isotopically traced species.

$$F_A = -\frac{d\mu_A^c}{dz} = -\frac{RT}{\bar{c}_A}\frac{d\bar{c}_A}{dz} \qquad\qquad\qquad\qquad (VII-2a)$$

and similarly for B:

$$F_B = -\frac{d\mu_B^c}{dz} = -\frac{RT}{\bar{c}_B}\frac{d\bar{c}_B}{dz}, \qquad\qquad\qquad (VII-2b)$$

where $\bar{c}_A$ is the gross concentration of A, in mol per cubic centimeter total volume.

Gross concentrations $\bar{c}_A$ and $\bar{c}_B$ are equal to the respective gas concentrations in free space times the porosity, ε (the porosity is the fraction of pore volume in the plug):

$$\bar{c}_A = \varepsilon c_A. \qquad\qquad\qquad\qquad\qquad\qquad (VII-3)$$

Generalized forces, F, concentration gradients, dc/dz, as well as fluxes J and superficial velocities u in later equations are vectors, the sign depending on the direction.

Although the physical nature of the exergy loss caused by this interdiffusion process is different from the salt and solvent transport phenomena in membranes, discussed in Chapter IV, it was shown by Stefan already in the 19th century that the friction model leads to phenomenological equations which are applicable here. One may consider the generalized driving force to the counterbalanced by a reaction (friction) force which consists of two terms: one due to momentum exchange in collisions between gas molecules diffusing in opposite directions, the other in collisions with the wall. The first friction term is proportional to the difference in the superficial velocities of the gases $u_A - u_B$ (cm s^{-1}) as in Eq. (IV-7). The other is proportional to the difference of the velocities of gas A and the matrix of the porous medium, C. The latter difference is imply u_A because the matrix is stationary. u_A and u_B have opposite signs in interdiffusion:

$$F_A = f_{AB}(u_A - u_B) + f_{AC}u_A, \qquad\qquad\qquad (VII-4)$$

where f_{AB} and f_{AC} are friction coefficients.[3] The porous matrix is the frame of reference.

---

3 It should be noted that, while in free space a simple relation exists between friction coefficients and mutual diffusion coefficients (which can often be found in tables), in porous media, the latter coefficients are replaced by "augmented diffusion coefficients" (Mason and Viehland 1978) which are not listed in conventional tables. With this reinterpretation, however, the frictional-model equations are correct.

The interaction between molecules A and B is caused by molecular collisions. Since the number of such collisions which 1 mol of A experiences per second is proportional to $\bar{c}_B$, we can set the friction factor, f_{AB}, proportional to $\bar{c}_B$, and define, in analogy with Eq. (IV-8), a mutual friction coefficient, β'', which is independent of the concentrations $\bar{c}_A$, $\bar{c}_B$ (but dependent on the geometry of the porous medium):

$$\beta'' \equiv \frac{f_{AB}}{\bar{c}_B}. \tag{VII-5}$$

Substituting for f_{AB} in Eq. (VII-4), we thus obtain

$$F_A = \beta'' \bar{c}_B (u_A - u_B) + f_{AC} u_A \tag{VII-6}$$

and similarly for F_B

$$F_B = \beta'' \bar{c}_A (u_B - u_A) + f_{BC} u_B. \tag{VII-7}$$

Rewriting Eqs. (VII-6) and (VII-7) in terms of the molar fluxes, $J_A = u_A \bar{c}_A$, $J_B = u_B \bar{c}_B$, we obtain

$$F_A = \left(\beta'' \frac{\bar{c}_B}{\bar{c}_A} + \frac{f_{AC}}{\bar{c}_A} \right) J_A - \beta'' J_B \tag{VII-8}$$

$$F_B = - \beta'' J_A + \left(\beta'' \frac{\bar{c}_A}{\bar{c}_B} + \frac{f_{BC}}{\bar{c}_B} \right) J_B. \tag{VII-9}$$

Comparing Eqs. (VII-8) and (VII-9) to Eqs. (IV-6), it is seen that they are of the same form, and that reciprocity of the resistance coefficients [Eq. (IV-18)], $R_{ij} = R_{ji} = (-\beta'')$ is satisfied.

Solving Eqs. (VII-8), (VII-9) for the fluxes, we obtain [4]

$$J_A = \underbrace{\frac{\bar{c}_A}{d''} (\beta'' \bar{c}_A + f_{BC})}_{L_{AA}} F_A + \underbrace{\beta'' \frac{\bar{c}_A \bar{c}_B}{d''}}_{L_{AB}} F_B \tag{VII-10}$$

---

[4] It is not surprising that in Eqs. (VII-10), (VII-11), reciprocity of the conductance coefficients [Eq. (IV-19)] is satisfied, because these two equations were derived by solution of the simultaneous Eqs. (VII-8), (VII-9), in which reciprocity ($R_{ij} = R_{ji}$) is valid.

$$J_B = \beta'' \frac{\bar{c}_A \bar{c}_B}{d''} F_A + \underbrace{\frac{\bar{c}_B}{d''} (\beta'' \bar{c}_B + f_{AC}) F_B}_{}, \qquad \text{(VII-11)}$$
$$\underbrace{\phantom{\beta'' \frac{\bar{c}_A \bar{c}_B}{d''}}}_{L_{BA}} \quad \underbrace{\phantom{\frac{\bar{c}_B}{d''} (\beta'' \bar{c}_B + f_{AC}) F_B}}_{L_{BB}}$$

where

$$d'' \equiv \beta'' (\bar{c}_A f_{AC} + \bar{c}_B f_{BC}) + f_{AC} f_{BC}. \qquad \text{(VII-12)}$$

These equations can be rewritten in the traditional form of diffusion equations [see remarks and footnote following Eqs. (IV-18), (IV-19)] by substituting for the generalized driving forces, F_A and F_B, from Eqs. (VII-2a, b)

$$J_A = -\underbrace{\frac{RT}{d''} (\beta'' \bar{c}_A + f_{BC})}_{D_{AA}} \frac{d\bar{c}_A}{dz} - \underbrace{RT\beta'' \frac{\bar{c}_A}{d''}}_{D_{AB}} \frac{d\bar{c}_B}{dz} \qquad \text{(VII-13)}$$

$$J_B = -\underbrace{RT\beta'' \frac{\bar{c}_B}{d''}}_{D_{BA}} \frac{d\bar{c}_A}{dz} - \underbrace{\frac{RT}{d''} (\beta'' \bar{c}_B + f_{AC})}_{D_{BB}} \frac{d\bar{c}_B}{dz}. \qquad \text{(VII-14)}$$

C. Diffusion Under Knudsen Conditions

If the gas pressure and/or the characteristic pore dimension is small, then the gas-wall collisions greatly outnumber the gas-gas collisions. Under these conditions, which are called Knudsen conditions, we can neglect the $A - B$ friction terms, characterized by β'', in Eqs. (VII-13) and (VII-14) in comparison with the gas-wall collision terms. If we designate the fluxes under these conditions (Knudsen fluxes) by J_{AK} and J_{BK}, respectively, Eqs. (VII-13) and (VII-14), respectively, yield:

$$J_{A,K} = -\frac{RT}{f_{AC}} \frac{d\bar{c}_A}{dz} = -\varepsilon \frac{RT}{f_{AC}} \frac{dc_A}{dz} \qquad \text{(VII-15)}$$

$$J_{B,K} = -\frac{RT}{f_{BC}} \frac{d\bar{c}_B}{dz} = -\varepsilon \frac{RT}{f_{BC}} \frac{dc_B}{dz}. \qquad \text{(VII-16)}$$

(Subscript K indicates Knudsen conditions.)

It is seen that under these conditions, each gas diffuses at the rate dictated by its own diffusivity, i.e., the fluxes are entirely uncoupled. The Knudsen diffusivities depend on the pore geometry, but not on the concentration of the gases, since f_{AC} and f_{BC} are independent of the gas concentrations (or partial pressures), in contrast to f_{AB} [Eq. (VII-5)]:

$$D_{A,K} = \varepsilon \frac{RT}{f_{AC}} \qquad\qquad\qquad\qquad\text{(VII-17)}$$

$$D_{B,K} = \varepsilon \frac{RT}{f_{BC}} . \qquad\qquad\qquad\qquad\text{(VII-18)}$$

The ratio of the Knudsen diffusion coefficients is:

$$\frac{D_{A,K}}{D_{B,K}} = \frac{f_{BC}}{f_{AC}} . \qquad\qquad\qquad\qquad\text{(VII-19)}$$

According to the kinetic theory of gases, this ratio equals the square root of the molecular mass ratio, m_B/m_A.

D. Interdiffusion at Uniform Pressure

1. Individual Rates of Diffusion

When two different ideal gases, originally at equal pressure and held in closed containers of equal volume, interdiffuse across a porous plug, the pressure does not necessarily remain uniform, because the gases can interdiffuse at different rates, thus building up a higher pressure in the container originally containing the gas of lower diffusivity.

It is possible to maintain uniform pressure, p, across a porous plug, however, even though the gases have different diffusivities, namely by replacing the two closed containers by two channels in which rapid flow of the gases A and B, respectively, both at the same pressure, is maintained. The total gas concentration in the porous plug, $\bar{c} = \bar{c}_A + \bar{c}_B$, is then uniform and

$$-\left(\frac{d\bar{c}_A}{dz}\right)_p = \left(\frac{d\bar{c}_B}{dz}\right)_p . \qquad\qquad\qquad\qquad\text{(VII-20)}$$

The gas fluxes are obtained from Eqs. (VII-13) and (VII-14) by substitution of $(d\bar{c}_B/dz) = -\varepsilon(dc_A/dz)$ from Eqs. (VII-20) and (VII-3):

$$J_A = -\frac{\varepsilon RT}{\beta''\left(\bar{c}_A \dfrac{f_{AC}}{f_{BC}} + \bar{c}_B\right) + f_{AC}}\left(\frac{dc_A}{dz}\right)_p \qquad\qquad\qquad\text{(VII-21)}$$

$$J_B = -\frac{\varepsilon RT}{\beta'' \left(\bar{c}_B \dfrac{f_{BC}}{f_{AC}} + \bar{c}_A \right) + f_{BC}} \left(\frac{dc_B}{dz} \right)_p .$$ (VII-22)

According to these equations, the flows obey the conventional diffusion behavior [Fick's equation, (III-26)], but the two species do not necessarily interdiffuse at the same rate. The two interdiffusion coefficients are

$$\bar{D}_A = \varepsilon RT \frac{f_{BC}}{d''}$$ (VII-23)

$$\bar{D}_B = \varepsilon RT \frac{f_{AC}}{d''} .$$ (VII-24)

It is of interest that the ratio of the interdiffusion coefficients at equal pressures is the same as the ratio of the Knudsen diffusion coefficients [Eqs. (VII-15) and (VII-16)], although the numerical values of each of the latter are larger $(D_{A,K} > \bar{D}_A; D_{B,K} > \bar{D}_B)$:

$$\frac{\bar{D}_A}{\bar{D}_B} = \frac{D_{A,K}}{D_{B,K}} = \frac{1/f_{AC}}{1/f_{BC}} .$$ (VII-25)

Many experiments under controlled conditions confirm this conclusion (Scott and Dullien 1962).

2. Equal Rates of Diffusion (Normal Interdiffusion)

It follows from Sect. D.1 that even though two interdiffusing gases are originally at equal pressure, a pressure difference will, in general, develop between the two containers (unless the pressure in each is regulated to remain constant – for instance, by continuous flushing), because of different rates of interdiffusion, characterized by *two different interdiffusion coefficients* [Eqs. (VII-23) and (VII-24)].

Because tables of diffusion coefficients list *only one interdiffusion coefficient*, it is pertinent to ask under which conditions the interdiffusing fluxes are numerically equal, remaining equal throughout the interdiffusion process, i.e., when

$$J_A = -J_B .$$ (VII-26)

Introducing this condition into Eq. (VII-8), we obtain

$$F_A = \frac{J_A}{\bar{c}_A} (\beta'' \bar{c} + f_{AC}) . \tag{VII-27}$$

$\bar{c}$ is the total gas concentration in the porous plug, which is uniform because the total gas pressure is uniform, although the ratio $\bar{c}_A / \bar{c}_B$ varies across the plug. Hence

$$\left(\frac{d\bar{c}_A}{dz} \right)_p = - \frac{J_A}{RT} (\beta'' \bar{c} + f_{AC}) . \tag{VII-28}$$

The analogous equation for gas B is

$$\left(\frac{d\bar{c}_B}{dz} \right)_p = - \frac{J_B}{RT} (\beta'' \bar{c} + f_{BC}) . \tag{VII-29}$$

These two expressions must be numerically equal and of opposite sign, because of the uniformity of the total pressure [Eq. (VII-21)]. Therefore, the two interdiffusing fluxes can be numerically equal [Eq. (VII-25)] only if

$$\beta'' \bar{c} + f_{AC} = \beta'' \bar{c} + f_{BC} . \tag{VII-30}$$

This condition is satisfied when $f_{AC} = f_{BC}$, i.e., when the Knudsen diffusion coefficients of the two gases are equal [Eqs. (VII-17) and (VII-18)]. This is the case for *self-diffusion* phenomena. In such cases, the transient pressure differences are very small, both the concentration gradients and the fluxes of the two gases are numerically equal and of opposite sign throughout the experiment, and the interdiffusion process is characterized by a single interdiffusion coefficient.

But from Eq. (VII-30) there is yet another situation in which both the concentration gradients and the diffusion fluxes between the closed vessels are numerically equal and opposed. This is the case when f_{AC} and f_{BC} are negligible, and even when the gases are not of similar nature. Physically, this means that the number of gas-wall collisions is negligible compared to the number of gas-gas collisions – i.e., the plug is highly permeable. This phenomenon is the opposite of Knudsen diffusion.

This condition is realized when the characteristic pore dimension is much larger than the mean free path (experiments at relatively high pressures, with relatively highly permeable plugs). In this case the total pressure remains constant, with respect to time, and uniform, and we obtain from Eqs. (VII-28) and (VII-29):

$$J_A = -J_B = -\frac{RT}{\beta''\bar{c}}\frac{d\bar{c}_A}{dz} = \frac{RT}{\beta''\bar{c}}\frac{d\bar{c}_B}{dz} = -\frac{RT}{\beta''c}\frac{dc_A}{dz} = \frac{RT}{\beta''c}\frac{dc_B}{dz}$$

$(p = \text{const})$. (VII-31)

This expression is of the type of Fick's law with a *single interdiffusion coefficient*, $\bar{D}_{AB}$:

$$\bar{D}_{AB} = \frac{RT}{\beta''c} = \varepsilon\frac{RT}{\beta''\bar{c}}.$$ (VII32)

Equation (VII-31) also describes interdiffusion in free space. In this case, one can consider the walls of the container in which the interdiffusion experiment is carried out as a porous medium consisting of a single pore. Here the gas-wall friction coefficients are entirely negligible. Equation (VII-32) then applies to any pair of gases at any pressure. Since, for ideal gases, β'' is independent of the composition (albeit not independent of the temperature and pore geometry), we see that for ideal gases there is only one interdiffusion coefficient which varies inversely with the concentration or total pressure.

Because those parts of Eqs. (VII-31) and (VII-32), which contain c, rather than $\bar{c}$, describe interdiffusion of gases in the absence of a specific porous medium, this type of diffusion is called *normal diffusion* or *normal interdiffusion*. (Isothermal) normal diffusion coefficients which, by definition, depend only on the properties of both components, are the diffusivities listed in physicochemical tables.

3. Pressure-Dependence of Self-Diffusion (Bosanquet Equation)

Consider the self-diffusion $A \leftrightarrow A^*$. This process takes place at uniform pressure. The fluxes of A and A^* are numerically equal and opposite in sign. Hence, Eq. (VII-27) applies, and the flux is

$$J_A = -J_B = -\frac{RT}{\beta''\bar{c} + f_{AC}}\frac{d\bar{c}_A}{dz} = -\varepsilon\frac{RT}{\beta''\bar{c} + f_{AC}}\frac{dc_A}{dz}.$$ (VII-33)

This equation is of the type of Fick's [Eq. (III-26)], the self-diffusivity being

$$\bar{D}_A^a = \varepsilon\frac{RT}{\beta''\bar{c} + f_{AC}}$$ (VII-34)

and its reciprocal is

$$\frac{1}{\bar{D}_A^a} = \frac{f_{AC}}{\varepsilon RT} + \frac{\beta''\bar{c}}{\varepsilon RT}.$$ (VII-35)

Comparison with Eqs. (VII-17) and (VII-32) shows that

$$\frac{1}{\bar{D}_A^a} = \frac{1}{D_{A,K}} + \frac{1}{\bar{D}_{AA^*}}. \tag{VII-36}$$

Since, for self-diffusion, $f_{AC} = f_{A^*C}$, Eq. (VII-34) confirms that the interdiffusion coefficients of the self-diffusing species are equal at any pressure. Since $1/\bar{D}_{AA^*}$ is proportional to the total gas concentration, $1/\bar{D}_A^a$ increases linearly with (but is not proportional to) the total pressure.

E. Summary

For a given gas pair, interdiffusion through a specific porous plug is characterized by a set of three parameters: the experimental ones $\bar{D}_A$, $\bar{D}_B$ (measured at $-dc_A/dz = dc_B/dz$), and $\bar{D}_{AB}$ (measured under conditions of "normal" interdiffusion, when $J_A = -J_B$); or the friction parameters f_{AC}, f_{BC}, and β'', which can be calculated from the three experimental coefficients by use of Eqs. (VII-17), (VII-18), and (VII-32), respectively; or the phenomenological coefficients L_{11}, L_{12}, and L_{22} of the general transport equations of non-equilibrium thermodynamics [Eqs. (IV-5)].

The description in terms of friction coefficients is useful because it leads to a number of important observed facts with a minimum of assumptions, and yet does not seem to contradict any observed findings. At least for ideal gases, this description also permits the calculation of interdiffusion at any desired pressure from data obtained at other pressures, which cannot be done directly by use of the L-coefficients. The phenomenological treatment is not meant to supersede more sophisticated traditional treatments in terms of the kinetic theory of gases which also contain the basic assumption about numerical equality of applied force and friction force (Newton's third law) expressed by Eq. (VII-4), and, for ideal gases, are based on the three parameters m_A, m_B (molecular masses), and the concentration of the dust representing the solid.

Selected Literature

Reviews of Kinetic Theory of Gases

Farkas A, Melville HW (1939) Experimental methods in gas reactions. Macmillan and Co, London
Knudsen M (1950) The kinetic theory of gases. 3rd edn, Methuen, London

Monographs on Transport in Porous Media

Dullien FAL (1979) Porous media: fluid transport and pore structure. Academic Press, New York
Carman PC (1956) Flow of gases through porous media. Academic Press, New York

Gas Diffusion in Porous Media

Evans III RB, Watson GM, Mason EA (1961) J Chem Phys 35:2076
Scott DS, Dullien FAL (1962) Am Inst Chem Eng J 8:113
Spiegler KS (1966) Ind Eng Chem (Fundamentals) 5:529
Mason EA, Viehland LA (1978) J Chem Phys 68:3562

Friction Model

Stefan J (1871, 1872) Wien Sitzber 63:(2) 63, 65:(2) 323
Spiegler KS (1966) See above

Gaseous Self-Diffusion and Bosanquet Equation

Pollard WG, Present RD (1948) Phys Rev 73:762

Chapter VIII. Molecular Filtration Through Membranes

A. Introduction

This chapter deals with the phenomenological and energetic aspects of molecular-filtration processes.[1] Advances in polymer chemistry have made it possible to prepare membranes which separate not only macromolecules and other colloids from solvents (*ultrafiltration*), but even molecular particles of comparable size (*hyperfiltration*; also called *reverse osmosis*). Large-scale hyperfiltration of sea water is possible, although the partial molal volumes of the salts and of water are of the same order of magnitude. Because of the importance of desalination and its widespread use, the example chosen in this chapter is the pressure-induced salt-water separation.

Nature's resistance against this demixing process manifests itself as the osmotic pressure, Π, i.e., the pressure which must be overcome to achieve complete separation with an ideally semipermeable membrane [defined as a membrane completely impermeable to solute (salt in this case) while being permeable to solvent (water in this case)].

B. Review of Osmotic-Pressure Equations[2]

Consider an ideally semipermeable membrane separating a salt solution from pure water (Fig. VIII-1).

If the pressures on *both* membrane faces were atmospheric, water would permeate the membrane from right to left, to decrease the salt concentration. To prevent this spontaneous process, the pressure, Π,[3] must be applied to the solution. In this state, there is no net water flow, and no net driving force for this flow, $F = -\Delta\mu_w$:

---

1 Note that the term *molecular sieves* is used for synthetic zeolites, which perform selective absorption in columns; these processes are not treated here.

2 From Spiegler KS (1977) Salt-Water Purification. 2nd edn, Plenum Press, New York and Spiegler and Kedem (1966)

3 Note that the osmotic pressure is a *property of the solution*. For the purpose of calculating the osmotic pressure of the solution, the membrane separating the solution from the pure solvent is always assumed to be ideally semipermeable.

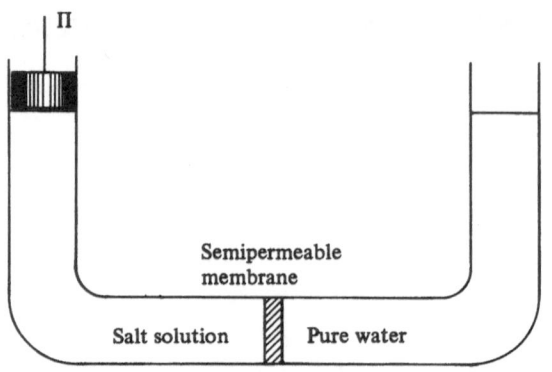

$$(\mu_w^{1a})^w = (\mu_w)^{sol}, \tag{VIII-1}$$

where $(\mu_w^{1a})^w$ is the chemical potential of pure water at 1 atm and $(\mu_w)^{sol}$ is the chemical potential of water in the solution at pressure Π.

The variation of the chemical potential with pressure is given by Poynting's equation:[4]

$$\left(\frac{\partial \mu_w}{\partial p}\right)_T = \bar{V}_w, \tag{VIII-2}$$

where $\bar{V}_w$ is the partial molal volume of water. Substituting this equation in Eq. (VIII-1), we obtain

$$(\mu_w)^{sol} = (\mu_w^{1a})^{sol} + \int_{1a}^{\Pi+1a} d(\mu_w)^{sol}$$

$$= (\mu_w^{1a})^{sol} + \int_{1a}^{\Pi+1a} \left(\frac{\partial \mu_w}{\partial p}\right)_T dp = (\mu_w^{1a})^{sol} + \bar{V}_w \cdot \Pi. \tag{VIII-3}$$

Hence, from Eq. (VIII-1), the osmotic pressure of the solution is

$$\Pi = \frac{(\mu_w)^{sol} - (\mu_w^{1a})^{sol}}{\bar{V}_w} = \frac{(\mu_w^{1a})^w - (\mu_w^{1a})^{sol}}{\bar{V}_w}. \tag{VIII-4}$$

If the membrane separates solutions, which differ in concentration by an infinitesimal amount, dc_s, the osmotic-pressure difference between the solutions is

$$d\Pi \equiv (\Pi)^{c_s+dc_s} - (\Pi)^{c_s} = -\frac{1}{\bar{V}_w}[(\mu_w^{1a})^{c_s+dc_s} - (\mu_w^{1a})^{c_s}] = -\frac{d(\mu_w^{1a})^{sol}}{\bar{V}_w}. \tag{VIII-5}$$

---

4 See, for instance, Hatsopoulos and Keenan 1965 (literature list of Chap. I) p 304.

To express the osmotic pressure in terms of the chemical potential of the *salt* rather than the water, we apply the Gibbs-Duhem equation[5]: for two solutions with infinitesimal concentration difference, the differences of the chemical potentials of salt and water are related to the respective mol fractions, y, by

$$\left(\frac{d\mu_s^{1a}}{d\mu_w^{1a}}\right)_T^{sol} = -\frac{y_w}{y_s} = -\frac{c_w}{c_s},$$
(VIII-6)

where y_w and y_s are the mol fractions which are proportional to the respective concentrations, c.

In dilute solutions, the partial molal volume of water (cm^3 water mol^{-1} water) is almost equal to the reciprocal of the water concentration (mol water cm^{-3} solution):

$$\bar{V}_w \cong 1/c_w.$$
(VIII-7)

Hence, we obtain from Eq. (VIII-6)

$$c_s d\mu_s^{1a} = -c_w d(\mu_w^{1a})^{sol} \cong -\frac{1}{\bar{V}_w} d(\mu_w^{1a})^{sol} \quad (p, T: const).$$
(VIII-8)

Comparison of this equation with Eq. (VIII-5) shows that

$$(d\Pi)_T = c_s d\mu_s^{1a}.$$
(VIII-9)

For ideal solutions of *electrolytes*, the chemical potential of the salt can be expressed by an equation equivalent to (III-8)

$$\mu_s^c = \mu_s^0 + \nu RT \ln c_s,$$
(VIII-10)

where ν is the total number of cations and anions produced by complete dissociation of one mol of electrolyte. For instance, for a $1-1$ electrolyte (e.g., NaCl), $\nu = 2$.

Hence, we obtain from Eqs. (VIII-9) and (VIII-10)

$$d\Pi = \nu RT dc_s \quad (ideal\ solution).$$
(VIII-11)

To find the osmotic pressure of a solution of concentration c_s, we consider a series of solutions of gradually increasing concentrations separated by mem-

---

5 See, for instance, Hatsopoulos and Keenan 1965, p 284, or Katchalsky and Curran 1965, p 54, Eq. (V-47). Both references are in the literature list of Chap. I.

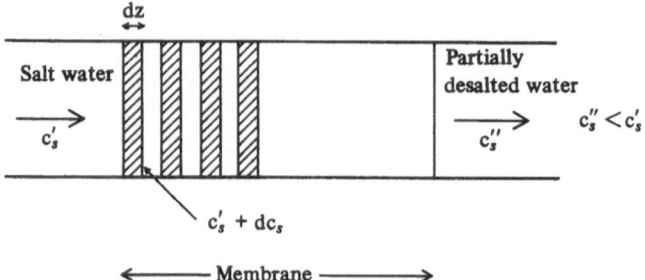

Fig. VIII-2. Schematic representation of hyperfiltration membrane. Membrane is broken down into differential elements, separated by uniform solution segments, which are in equilibrium with the two contiguous membrane faces. Spiegler and Kedem (1966) Desalination 1:133

branes, as shown in Fig. VIII-2, and integrate from $c_s = 0$ (pure solvent) to c_s (in this case, the membranes are assumed to be ideally semipermeable):

$$\int_0^\pi d\varPi \equiv \varPi = \nu RT \int_0^{c_s} dc_s = \nu RT c_s. \qquad \text{(VIII-12)}$$

It is convenient to characterize the deviation of a solution from ideal behavior, by the ratio of its osmotic pressure to the osmotic pressure calculated from Eq. (VIII-12):

$$\varPi = g\nu RT c_s. \qquad \text{(VIII-13)}$$

This ratio, g, is called the *rational osmotic coefficient*. It varies with the concentration, c_s.

In systems with concentration gradients, such as the one shown in Fig. VIII-2, it is often useful to calculate the *gradient of the osmotic pressure*. Differentiation of Eq. (VIII-13) yields

$$\frac{d\varPi}{dz} = \nu RT \left(g \frac{dc_s}{dz} + c_s \frac{dg}{dz} \right) = \nu RT g \left(1 + \frac{d \ln g}{d \ln c_s} \right) \frac{dc_s}{dz}. \qquad \text{(VIII-14)}$$

In hyperfiltration through ideally semipermeable membranes, pressure *larger* than the osmotic pressure of the solution (Fig. VIII-1) is applied to the solution, so as to force solvent through the membrane from left to right. Thus, for ideally semipermeable membranes, $\varPi$ represents a watershed value. If the pressure applied on the left is smaller than $\varPi$, solvent moves from right to left (osmosis); if the pressure is higher than $\varPi$, solvent moves from left to right (reverse osmosis). The membranes available to industry are rarely ideally semipermeable, however. The flow of solute must also be taken into account, as shown in Sect. C.

C. Flow Equations[6]

Since practical molecular-filtration membranes are not ideally semipermeable, complete solvent-solute separation in a single passage of the fluid through the membrane is, in general, not possible. This section deals with the dependence of the degree of separation on the production rate and, hence, the exergy dissipated by the separation process, as well as the membrane parameters which determine this rate and the degree of separation which can be achieved.

The practical molecular-filtration equations can be derived from flow equations of the type of (V-1). It is possible, however, to make these equations plausible by using the simple mosaic membrane model of Fig. VIII-3. In this model, the membrane is composed of two regions of radically different properties, viz., an ideally semipermeable region (area fraction y_a), which passes water but no salt, and an entirely non-selective region (area fraction y_b) through which the high-pressure solution passes unchanged. The hydraulic permeabilities of the

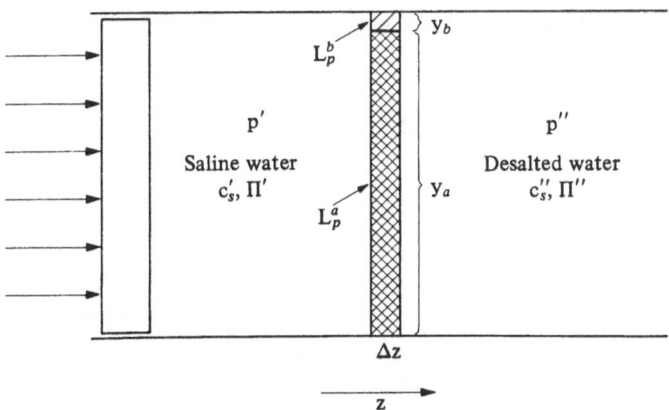

Fig. VIII-3. Schematic of desalination by mosaic membrane. The figure shows unit area of membrane. Region a of area y_a and hydraulic permeability L_p^a is ideally semipermeable (i. e., impermeable to salt); region b of area y_b and hydraulic permeability L_p^b is completely unselective. Harris, Humphreys and Spiegler (1976). Reverse osmosis in water desalination Chap. IV in: Meares P (ed) Membrane separation processes. Elsevier, Amsterdam 1976

---

6 This section follows the development of Sect. I.1.D of the author's joint chapter with Harris FL and Humphreys GB (1976) In: Meares P (ed) Membrane Separation Processes. Elsevier Amsterdam.

two regions are L_p^a and L_p^b, respectively. (In reality, imperfect membranes do not necessarily consists of perfect regions a, and pin-holes, b, but for the purpose of illustrating the flow equations, it is didactically useful to introduce such a model.) We consider the application of pressure to the salt solution on the left (salt concentration c_s'), which results in production of the hyperfiltrate on the right (salt concentration c_s''). $c_s'' - c_s' = \Delta c_s < 0$.

In Fig. VIII-3, the driving force through the ideally semipermeable portion of the membrane, a, is $-(\Delta p - \Delta \Pi)$, while the driving force through portion b is $-\Delta p$, by Darcy's law. Accordingly, the total volume flux through the mosaic membrane is given by

$$J_v = - [L_p^a y_a (\Delta p - \Delta \Pi) + L_p^b y_b \Delta p] = - L_p(\Delta p - \sigma \Delta \Pi), \qquad \text{(VIII-15)}$$

where

$$L_p = L_p^a y_a + L_p^b y_b \qquad \text{(VIII-16)}$$

and

$$\sigma = \frac{L_p^a y_a}{L_p^a y_a + L_p^b y_b}. \qquad \text{(VIII-17)}$$

σ is the *reflection coefficient* of the membrane. The physical meaning of the reflection coefficient can be seen when the equation is applied to a membrane which is rapidly flushed on both faces with the same solution. In this case, $\Delta \Pi$ is zero and Eq. (VIII-15) reduces to

$$(J_v)_{\Delta \Pi = 0} \simeq - (L_p^a y_a + L_p^b y_b) \Delta p = - L_p \Delta p. \qquad \text{(VIII-18)}$$

Hence,

$$\sigma = L_p^a y_a / L_p; (1 - \sigma) = L_p^b y_b / L_p. \qquad \text{(VIII-19)}$$

It is seen that, for $\Delta \Pi = 0$, σ represents the fractional contribution of the ideal part of the membrane to the total permeability, L_p.

Salt can be transported across the mosaic membrane only in the element b, which represents a macroscopic pore. Salt transport in such a pore takes place by diffusion and convection; also, the local velocities of salt migration caused by these two processes are additive. Although this does not imply that the integral diffusion and convection velocities or the superficial fluxes through a membrane of finite thickness Δz are necessarily additive also, this approximation is frequently made and used here.[7] Thus

---

7 For a discussion of the limitations of this assumption, the reader is referred to Helfferich 1962 (pp 332 – 333).

$$J_s = -y^b \bar{D}_s^b (\Delta c_s / \Delta z) + c_s' J_v^b \tag{VIII-20}$$

where $\bar{D}_s^b$ is the diffusion coefficient in region b and J_v^b the volume flow through b. When $\Delta p \gg \Delta \Pi$, Eq. (VIII-18) shows that the term $-L_p^b y_b \Delta p$ represents the volume flow J_v^b through region b. On expressing J_v^b in terms of the reflection coefficient, σ, from Eq. (VIII-19), we obtain

$$J_v^b = J_v L_p^b y_b / L_p = (1 - \sigma) J_v . \tag{VIII-21}$$

Introducing this value for J_v^b into Eq. (VIII-20) and defining the solute permeability:

$$\omega \equiv y^b \bar{D}_s^b / (\nu RT \Delta z) \tag{VIII-22}$$

we obtain

$$J_s = -\omega \Delta \Pi + (1 - \sigma) J_v c_s' \simeq -\nu \omega RT \Delta c_s + (1 - \sigma) J_v c_s' . \tag{VIII-23}$$

At low pressures, when J_v is low, the first term on the right of the salt flux equation, Eq. (VIII-20), is dominant. This pressure region is of no practical importance in hyperfiltration. At high pressures the second term is dominant and it follows from Eq. (VIII-21) that

$$J_s = c_s' (1 - \sigma) J_v \quad (|\Delta p| \gg |\Delta \Pi|) . \tag{VIII-24}$$

In the steady state, the hyperfiltrate concentration c_s'' equals the ratio of the salt and volume fluxes. Therefore

$$c_s'' = J_s / J_v = c_s' (1 - \sigma) \quad (|\Delta p| \gg |\Delta \Pi|) . \tag{VIII-25}$$

The hyperfiltration ability of a membrane with respect to a given salt is often listed as the ratio of the decrease in salt concentration and the bulk salt concentration of the saline water. This ratio is called the *salt rejection coefficient*, **R**, where

$$\mathbf{R} \equiv (c_s' - c_s'')/c_s' = 1 - (c_s''/c_s'); \quad c_s'' = c_s' (1 - \mathbf{R}) . \tag{VIII-26}$$

On comparing Eqs. (VIII-25) and (VIII-26), it is seen that, at high pressures, **R** should equal the reflection coefficient σ [Eq. (VIII-15)]:

$$\sigma = (\Delta p / \Delta \Pi)_{J_v = 0} . \tag{VIII-27}$$

It is of interest to predict the degree of separation, not only at high pressures [Eqs. (VIII-24), (VIII-25)], but to represent the salt-rejection coefficient as a function of applied pressure when $|\Delta p| < |\Delta \Pi|$. For this purpose, the basic flow equations for volume flux (VIII-15) and salt flux (VIII-23) are first transformed into differential form, by applying them to a membrane lamella of infinitesimal thickness (Fig. VIII-2):

$$J_v = -\mathscr{P}_1\left(\frac{dp}{dz} - \sigma\frac{d\Pi}{dz}\right) \tag{VIII-28}$$

$$J_s = -\bar{P}\frac{dc_s}{dz} + (1-\sigma)c_s J_v. \tag{VIII-29}$$

Here, p, c and Π refer to the average parameters measured in the solutions between the lamellae. The membrane permeability parameters, L_p and ω, respectively, have been normalized for unit membrane thickness:

$$L_p \equiv \mathscr{P}_1/\Delta z \tag{VIII-30}$$

and

$$\omega = \frac{\bar{P}}{\nu RT(\Delta z)}. \tag{VIII-31}$$

Integration of Eq. (VIII-29), based on the assumption of constant, concentration-independent values of the parameters σ and $\bar{P}$, leads to a quantitative expression for the change of salt rejection with product flux.[8] When the applied pressure is increased from zero to values larger than $\Delta \Pi$, the salt rejection rises rapidly at first and then reaches a limiting value, σ, asymptotically.

The difference between perfectly semipermeable and imperfect membranes is therefore: with perfect membranes, the product is always pure water and driving pressures higher than the osmotic pressure of the solution have to be applied; with imperfect membranes pressure differences much smaller than the osmotic pressure of the raw water may cause hyperfiltrate flow but the quality of the product is poor under these circumstances (small **R**). The salt rejection increases as the pressure is increased and reaches a limiting value. These predictions from the theory have been borne out by ultrafiltration and hyperfiltration of different solutions.

Three practical parameters, L_p, σ, and ω, are required to characterize each (membrane + solutions) system. The necessity for three parameters appears in the thermodynamic treatment which is based on three conductivity coefficients,

---

8 For details, the reader is referred to the paper by Spiegler and Kedem (1966).

L_{ij}. In practice, some correlation between ω and σ is to be expected, however. For instance, very "leaky" membranes (low σ) may be expected to have more pathways for salt diffusion (large ω) than have tight membranes and, ceteris paribus, higher hydraulic permeabilities L_p also. However, L_p and ω both depend strongly on the thickness of the selective membrane and correlations between the practical parameters are valid only when the results are normalized to hyperfiltration layers of the same thickness. Moreover, σ, ω and, to a lesser extent L_p, depend strongly on the nature of the solute; thus a membrane may act as an effective hyperfilter for one solute (e. g., Na_2SO_4) while being leaky with respect to another (e. g., NaCl). This difference is reflected in the values of the practical coefficients for the different salts. Note also that many practical hyperfilters consist of more than one layer. For instance, modified cellulose acetate, used frequently in desalination, has a thin, dense skin, supported by a more porous base layer. This structure can give rise to complex salt-concentration patterns within the membrane.

Problems

VIII.1. The suggestion has been made to build a device which produces fresh water from sea water by lowering into the ocean a pipe filled with fresh water and closed at its end by a semipermeable membrane (i. e., a membrane permeable to water, but not to salts). If the pipe is lowered to a sufficient depth, the hydrostatic pressure on the outside of the pipe exceeds the hydrostatic pressure on the inside by more than the osmotic pressure of sea water. It was claimed that fresh water would now flow through the membrane into the pipe and that fresh water could then be collected continuously at the top of the pipe (Scientific American Dec 1971 p 100).

a) Given that the vapor pressure of sea water is 1.84% less than that of pure water, calculate the exergy (with respect to a reservoir at 25 °C) necessary to produce 1 ton of pure water from a large amount of sea water at 25 °C. Express the answer in kWh tonne^{-1}.

b) Calculate the osmotic pressure of sea water.

c) Calculate the ocean depth at which the difference between the hydrostatic pressures of sea water and fresh water equals the osmotic pressure of sea water. Assume that the average density and temperature of sea water in the location considered are 1.026 g cm^{-3} and 25 °C, respectively, and neglect the variation of the temperature with depth.

d) Discuss whether, *in principle*, the proposed scheme could work (if the columns of sea water and fresh water are assumed to be at rest).

Given: Partial pressure of pure water at 25 °C: 0.0312 atm. Specific volume of this water vapor: 43.4 l g^{-1}. Density of pure water at 25 °C: 0.997 g cm^{-3}. Acceleration of gravity: $g = 980.7$ cm s^{-2}.

VIII.2.

a) Certain membranes are too small and/or fragile for direct measurement of their hydraulic permeability. The suggestion has been made to make an osmotic measurement instead, and calculate the hydraulic permeability therefrom (i.e., place a solution on one side and pure solvent on the other, and then measure the total volume flux as a function of the solute concentration). State for what kind of membrane-solution systems, if at all, this procedure is permissible. What conditions must the transport coefficients satisfy?

b) A further suggestion has been made to substitute water self-diffusion experiments for osmotic permeability experiments, by equating the hydraulic or osmotic mobility of water to the self-diffusion mobility. The following experimental results were obtained for different membranes. (In the osmotic experiments, solutes were used to which the membranes are impermeable.)

Membrane	$g_{do} \equiv$ Osmotic mobility/ Diffusion mobility[9]
Cellophane	80
Polyvinyl alcohol film	12.5
Polyvinyl chloride soaked with tributylphosphate	1.8

(Tributylphosphate is a liquid which is almost insoluble in water. The last membrane may thus be considered a liquid membrane.)

Calculate g_{do} from the friction model of membrane transport processes, expressing it in terms of the molar concentration of water in the membrane and the Rayleigh-Einstein friction coefficients. From this expression, prove that $g_{do} \geq 1$, as observed experimentally. State the conditions for which $g_{do} = 1$ in terms of transport coefficients for the system.

Hint: Note that the mobility of a species is defined as the flux divided by the concentration and the generalized driving force acting on this species (Chap. III).

c) Under what conditions is it possible to calculate the hydraulic permeability of porous materials to *gas* flow from gas interdiffusion experiments? What is the formal analogy between the transport phenomena in the liquid phase [part b) of this problem] and the gas transport case? Discuss Knudsen and normal gas diffusion from this viewpoint.

---

9 Thau et al. 1966.

Selected Literature

Reviews

Lonsdale HK (1982) The growth of membrane technology. J Membr Sci 10:81
Pusch W, Walch A (1982) Membrane structure and its correlation with membrane permeability. J Membr Sci 10:325
Soltanieh M, Gill WN (1981) Review of reverse-osmosis membranes and transport models. Chem Eng Comm 12:279
Dresner L, Johnson JS Jr (1980) Hyperfiltration (reverse osmosis). In: Spiegler KS, Laird ADK (eds) Principles of desalination. 2nd edn, Academic Press, New York
Sourirajan S (1977) Reverse osmosis. Nat Res Counc Can Ottawa
Harris FL, Humphreys GB, Spiegler KS (1976) Reverse osmosis (hyperfiltration) in water desalination. In: Meares P (ed) Membrane separation processes. Elsevier Amsterdam
Merten U (1966) Desalination by reverse osmosis. MIT Press, Cambridge Mass

Non-Equilibrium Thermodynamics Applied to Hyperfiltration

Johnson JS Jr, Dresner L, Kraus KA (1966) In: Spiegler KS (ed) Principles of desalination. 1st edn, Academic Press, New York
Spiegler KS, Kedem O (1966) Thermodynamics of hyperfiltration (reverse osmosis). Desalination 1:311
Katchalsky A, Curran P (1965) see literature list of Chapter I
Kedem O, Katchalsky A (1958) Thermodynamic analysis of the permeability of biological membranes to non-electrolytes. Biochim Biophys Acta 27:229
Staverman AJ (1952) Non-equilibrium thermodynamics of membrane processes. Trans Faraday Soc 48:176
Staverman AJ (1951) The theory of measurement of osmotic pressure. Rec Trav Chim 70:344

Other References Relevant to Chapter VIII

Spiegler KS (1977) Salt-water purification, 2nd edn, Ch 9, Plenum Press, New York
Levenspiel O, de Nevers N (1974) The osmotic pump. Science 183:157
Bean CP (1972) The physics of porous membranes-neutral pores. In: Eisenman G (ed) Membranes. Vol I. Marcel Dekker, New York
Gröpl R, Pusch W (1970) Asymmetric behavior of cellulose-acetate membranes. Desalination 8:277
Thau G, Bloch R, Kedem O (1966) Water transport in porous and non-porous membranes. Desalination 1:129
Helfferich FG (1962) Ion exchange. McGraw-Hill, New York

Chapter IX. Coupled Heat and Mass Flow

A. Introduction

Many effects due to the coupling of heat with mass flow, e. g., thermomechanical effects and thermomolecular pressure, have been studied extensively, and some, such as separations by thermal diffusion, have been applied in practice. This chapter is an introduction to the fundamentals of the coupling of heat flow with ideal-gas flow, introducing some basic energy concepts associated with these coupled-flow phenomena, e. g., the *heat of transport*.[1] I also believe that an understanding of these phenomena is helpful in conceptualizing phenomenologically the laws of *thermoelectricity* (discussed in Chap. X) which are of major importance in energy technology.

B. Adiabatic Gas Flow Through Porous Media

1. Joule-Thomson Effect

Consider the thermally insulated system shown in Fig. IX-1, consisting of a porous diaphragm separating two spaces filled with the same gas. This gas is made to flow through the diaphragm from left to right under the influence of pressure. It is of interest to determine what, if any, temperature change occurs in the gas upon passage through the diaphragm.

In Chapter VII it was shown that the physical nature of gas transport in porous media depends on the ratio of the characteristic pore dimension to the mean free path of the gas. Therefore, the question about the temperature change cannot be answered unless additional information is available on *both* the gas and the porous plug. When the characteristic pore size is much larger than the mean free path, then a normal statistical distribution of molecular energies, determined only by the initial gas temperature and pressure, prevails at any location in the plug, because the gas-gas collisions which determine this distribution greatly outnumber the gas-wall collisions. For instance, when an ideal gas passes

---

1 Also called *heat of transfer.*

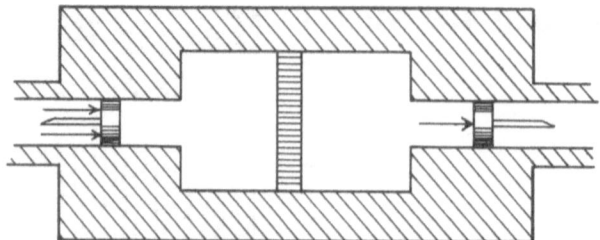

Fig. IX-1. Joule-Thomson effect

through such a diaphragm (which is assumed to be chemically inert), one cannot expect the diaphragm to be thermoselective, i.e., to alter the temperature by altering the initial statistical distribution of the molecular energies.

Whenever the characteristic pore dimension is much larger than the mean free path of the permeating (ideal or non-ideal) gas, the permeation process is called *Joule-Thomson effect*.

Because the system shown in Fig. IX-1 is thermally insulated, Eq. (I-6), applied to 1 mol of gas permeating the porous diaphragm, reduces to

$$Q = 0 = \Delta\bar{U} + W, \tag{IX-1}$$

where the overbar indicates the molar quantity.

Hence:

$$\Delta\bar{U} = -W = p''\bar{V}'' - p'\bar{V}' \tag{IX-2}$$

or

$$\bar{U}'' + p''\bar{V}'' = \bar{U}' + p'\bar{V}'. \tag{IX-3}$$

It is seen that the permeation process is isenthalpic [see also Sect. II.C.3, Eq. (II-16):

$$\bar{H}'' = \bar{H}'. \tag{IX-4}$$

If the permeating gas is ideal, this implies that no temperature change occurs, because for an ideal gas the enthalpy increases with the temperature, while being independent of the pressure. If the gas is not ideal, temperature changes generally do occur. The Joule-Thomson coefficient, K_{JT}, defined as

$$K_{JT} \equiv \left(\frac{\partial T}{\partial p}\right)_H \tag{IX-5}$$

can be negative or positive, depending on the nature and state of the gas.

2. Thermoselective Diaphragms

When the characteristic pore dimension is much smaller than the mean free path of the gas molecules (Knudsen conditions, as discussed in Chap. VII), the flows of different types of molecules are not completely coupled. It is even possible that the permeating species emerges from the diaphragm with a changed energy distribution.

In principle, such a thermoselective separator can retard the passage of the less, or of the more energetic molecules. Assume the temperature in the right space gets lower.[2] In Fig. IX-2, permeation of an ideal gas through a thermoselective separator at uniform temperature is schematically described. If the permeation process were to take place adiabatically, the temperature in the right and left compartments would decrease and increase, respectively. To maintain the temperature uniform, it is necessary to cool and heat the two compartments, respectively, i. e., to maintain a heat flow, opposed to the mass flow, through the plug. Thus, the maintenance of constant temperature implies a coupling of mass and heat flow in opposite directions.

C. Flux Equations

1. Heat of Transfer

The flux equations for the coupled mass and heat fluxes, named the *thermomechanical effect*, are of the form of Eq. (V-2). Fluxes, J, and generalized driving forces are conjugated as listed in Table III-1, and the flux equations written in analogy to the coupled-flux equations for electroosmosis (IV-44) and (IV-45)[3] are:

$$J_k = L_{kk}\left(-\frac{\bar{V}_k}{T}\frac{dp}{dz}\right) + L_{kQ}\left(-\frac{1}{T^2}\frac{dT}{dz}\right) \qquad \text{(IX-6)}$$

---

2 For ideal gases, permeating a porous diaphragm under Knudsen conditions, experimental evidence indicates that this is so. (Experiments by M. Knudsen on molecular flow of hydrogen or of air through a dense plug established that the gas temperature at the exit was lower, as reported by Sophus Weber (1924) Z Phys 24:267; Fig. 1, p 272.) This conclusion can also be drawn from the fact that the heat of transfer for ideal gases is negative, as shown in Sect. IX.D, Eq. (IX-24). The heat of transfer is the ratio of the heat flow, which must be maintained to keep the temperature uniform, to the gas flow. An analogous phenomenon in *condensed phases* is known as the *Dufour effect*

3 Here the generalized driving force for permeating under pressure has been multiplied by the partial molal volume of the gas, V_k, as explained in Sect. IV.B.2.

The units of L_{kQ} and L_{Qk} are mol cm^{-1} s^{-1} deg, and of L_{QQ} and L_{kk} W cm^{-1} deg and W^{-1} cm^{-1} s^{-2} mol^2 deg, respectively.

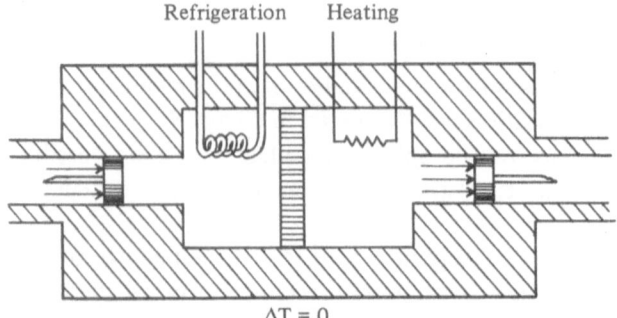

Fig. IX-2. Thermomechanical effect

$$J_Q = L_{Qk}\left(-\frac{\bar{V}_k}{T}\frac{dp}{dz}\right) + L_{QQ}\left(-\frac{1}{T^2}\frac{dT}{dz}\right). \tag{IX-7}$$

In the linear range, these equations are valid for any value of the driving forces, even for the case depicted in Fig. IX-2, in which the temperature gradient is zero. In this situation, the ratio of heat to mass flow is

$$\left(\frac{J_Q}{J_k}\right)_T = \frac{L_{Qk}}{L_{kk}} \equiv Q^*. \tag{IX-8}$$

This ratio (W s mol^{-1}) is called the *heat of transfer.*

For an ideal gas, uniformity of temperature implies uniformity of enthalpy in steady-flow (Fig. IX-2). This means that the heat added to the right container by the heater is numercially equal to the heat removed from the left container by the refrigerator. For fluids other than ideal gases, this is not necessarily so. It is (the absolute value of) the heat flow out of the high-pressure container (left in Fig. IX-2) which is listed as the heat of transfer in the literature.

2. Thermomolecular Pressure Difference

When two vessels containing the same gas, and separated by a thermoselective diaphragm, are brought to different temperatures, a pressure difference develops across the diaphragm (Fig. IX-3). This *thermomolecular pressure difference* is phenomenologically analogous to the electroosmotic counterpressure (Sect. IV.D) and is a consequence of the coupling between heat and mass flow described by Eqs. (IX-6) and (IX-7). In the stationary state, the gas flow due to the pressure difference between the two vessels is counterbalanced by the gas flow (in the opposite direction) caused by coupling to the heat flow, so that the net flow, J_v, vanishes.

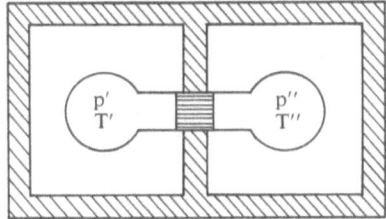

Fig. IX-3. Thermomolecular pressure difference. The two gas containers are separated by a thermoselective porous medium. When the gas temperatures in the containers are different, so are the pressures in the steady state (no net gas transfer). For ideal gases, $\Delta p/\Delta T > 0$

In this condition, Eq. (IX-6) reduces to

$$\left(\frac{dp}{dT}\right)_{J_k=0} = -\frac{1}{\bar{V}_k T}\frac{L_{kQ}}{L_{kk}}. \tag{IX-9}$$

(It was assumed that the contribution of the solid skeleton of the diaphragm to the heat flow, J_Q, is negligible.)

Introducing the heat of transfer from Eq. (X-8), we obtain

$$\left(\frac{dp}{dT}\right)_{J_k=0} = -\frac{1}{\bar{V}_k T}Q^* \equiv K_K, \tag{IX-10}$$

where K_K is the coefficient of thermomolecular pressure.

Note that for ideal gases, Q^* is negative (Sect. IX.D). Hence, K_K is positive. This means that when no net gas flow takes place, the gas in the container of higher temperature has the higher pressure.

Thermomolecular pressure differences can also develop in liquids (Soret effect, or *thermoosmosis*) and aqueous membranes (Pagliuca 1983). This has been observed even in liquid helium below 219 K (*fountain effect*). For gases, the effect is often called the *Knudsen effect*. It can only come into play when the separator is thermoselective.

D. Heat of Transfer of Ideal Gases (Gas-Kinetic Interpretation)[4]

When an ideal gas flows isothermally through a porous diaphragm the characteristic pore dimension of which is much larger than the mean free path of the gas molecules (Fig. IX-1), then each mol carries the energy $\bar{e}^*_{JT}$:

$$\bar{e}^*_{JT} \equiv \bar{H}. \tag{IX-11}$$

---

4 Knudsen 1950, Prigogine 1967.

If the diaphragm is now replaced by one having small pores, or if the gas pressure is greatly reduced (characteristic pore dimension ≪ mean free path), the previous Joule-Thomson flow changes into Knudsen flow i.e., the diaphragm has become thermoselective. Under these conditions, the energy transferred by one mol gas is

$$\bar{e}_k^* = \bar{H} + Q^*. \tag{IX-12}$$

It is possible to calculate the heat of transfer, Q^*, by elementary considerations of the kinetic theory of gases and show that $Q^* < 0$.

Consider the stationary state in a thermomolecular-pressure experiment (Fig. IX-3). The pressure exerted on each cm^2 of the left side of the diaphragm is the product of the number of the impacts per sec,[5] $\dot{N}'$, with the momentum change per molecule $2m\bar{u}f_M$, where m is the molecular mass (g molecule^{-1}), $\bar{u}$ the (arithmetical) mean molecular velocity, and f_M (<1) a geometrical factor which corrects for the fact that not all impacts are perpendicular (f_M is independent of pressure and temperature). Hence, the pressure exerted on the left side of the diaphragm is

$$p' = \dot{N}'(2m\bar{u}f_M). \tag{IX-13}$$

Hence,

$$\dot{N}' = p' \bigg/ \left(2mf_M\sqrt{\frac{8kT'}{\pi m}}\right) = p' \bigg/ \left(4f_M\sqrt{\frac{2mk}{\pi}T'}\right). \tag{IX-14}$$

Not all the molecules impacting on one face of the diaphragm eventually emerge at the other face, since some are reflected. Assume that a fixed fraction, f_G, depending only on the geometry of the porous diaphragm of the number impacting, moves across the diaphragm. In the stationary state (Fig. IX-3), the number of molecules crossing from left to right equals the number crossing from right to left:

$$(\dot{N}_{\rightarrow})_{J_k=0} = \dot{N}'f_G = (\dot{N}_{\leftarrow})_{J_k=0} = \dot{N}''f_G. \tag{IX-15}$$

Here, single and double primes designate the left and right compartments, respectively.

Substituting for $\dot{N}'$ from Eq. (IX-14), and using an analogous expression for $\dot{N}''$, we obtain from Eq. (IX-15)

---

5 According to the kinetic theory of gases (see, for instance, Knudsen 1950, p 3), this number is $(n/4)\bar{u}$, where n is the molecular concentration (molecules cm^{-3}), and the mean velocity is $\bar{u} = [8kT'/(\pi m)]^{1/2}$, where k is Boltzmann's constant (1.38×10^{-23} W s molecule^{-1} deg^{-1}).

$$\left[p' f_G \Big/ \left(4 f_M \sqrt{\frac{2 mk}{\pi} T'}\right)\right]_{J_k = 0} = \left[p'' f_G \Big/ \left(4 f_M \sqrt{\frac{2 mk}{\pi} T''}\right)\right]_{J_k = 0}.$$

(IX-16)

Hence,

$$\left(\frac{p'}{p''}\right)_{J_k = 0} = \left(\sqrt{\frac{T'}{T''}}\right)_{J_k = 0}$$

(IX-17)

known as the *Knudsen equation*.

It is seen that, if the temperatures of the gases in the containers shown in Fig. IX-3 are systematically changed, the stationary-state pressures which will eventually established themselves are such as to conform to the Knudsen equation. Comparing the stationary states at different temperatures, the expression $\left(\dfrac{p}{\sqrt{T}}\right)_{J_k = 0}$ is the same constant, K, for the gases on both sides of the diaphragm. Hence,

$$(p)_{J_k = 0} = (K T^{1/2})_{J_k = 0}.$$

(IX-18)

The change of the stationary pressure, p, in a given compartment with the temperature, T, in that compartment (while the temperature and pressure in the other compartment are constant) is obtained by differentiation:

$$\left(\frac{dp}{dT}\right)_{J_k = 0} = \frac{1}{2} K T^{-1/2}.$$

(IX-19)

We substitute the value of K from Eq. (IX-18):

$$\left(\frac{dp}{dT}\right)_{J_k = 0} = \frac{1}{2} \frac{p}{\sqrt{T}} T^{-1/2} = \frac{1}{2} \frac{p}{T}.$$

(IX-20)

Introducing the ideal-gas law

$$p = RT/\bar{V}_k$$

(IX-21)

we obtain

$$\left(\frac{dp}{dT}\right)_{J_k = 0} = \frac{1}{2} \frac{R}{\bar{V}_k}$$

(IX-22)

or, for finite pressure and temperature differences between the containers:

$$\left(\frac{\Delta p}{\Delta T}\right)_{J_k=0} = \frac{1}{2}\frac{R}{\bar{V}_{k,m}}, \tag{IX-23}$$

where $\bar{V}_{k,m}$ is a suitable average of the molar volumes of gas k in the two containers.

Comparing Eq. (IX-22) to (IX-10), we see that the heat of transfer for an ideal gas is

$$Q^* = -\frac{1}{2}RT. \tag{IX-24}$$

It is seen that this heat of transfer is negative. According to Eq. (IX-8), this implies that the heat flow which must be produced in a thermomechanical-effect experiment with an ideal gas (Fig. IX-2) so as to keep the temperature uniform opposes the mass flow (negative coupling), as was indeed assumed in the first place.[6]

E. Summary

The energy transported per mol, when an ideal gas flows across a diaphragm under Joule-Thomson conditions (Sect. IX.B.1) is

$$\bar{e}^*_{JT} = \bar{H}. \tag{IX-11}$$

For a *monatomic* ideal gas [the internal energy of which is $(3/2)RT$], this amounts to

$$\bar{e}^*_{JT} = \bar{U} + p\bar{V}_m = \bar{H} = \tfrac{3}{2}RT + RT = \tfrac{5}{2}RT. \tag{IX-25}$$

On the other hand, if the flow occurs under Knudsen conditions, the energy transported per mol is

$$\bar{e}^*_K = \bar{H} + Q^*, \tag{IX-12}$$

which is smaller than the energy transported under Joule-Thomson conditions [Eq. (IX-25)], because for ideal gases, the heat of transfer, Q^*, is negative.

---

6 The reasoning leading to Eq. (IX-24) in Sect. IX.D in no way depends on this
  assumption and the discussions in Sect. IX.B.2 and IX.C, however.

For monatomic ideal gases

$$\bar{e}_K^* = 2\,RT\,. \tag{IX-26}$$

Problem

IX.1.
a) Vessels I and II, containing oxygen, are connected by a thermoselective plug. The temperatures in I and II are maintained at 370 and 382 K, respectively, and the pressure in Vessel I in the steady state is one atm. What is the pressure in Vessel II?
b) A steady flow of 4245 l min^{-1} of an ideal gas at atmospheric pressure is maintained through this system. What must be done to keep the temperature in both vessels at 376 K? If heating and refrigeration are necessary, state where and calculate the heat flows per minute (in kcal min^{-1}).
Molar volume of ideal gas at 1 atm, 0°C = 22.3 l mol^{-1}.
1 kcal = 41.31 l atmosphere.

Selected Literature

General

Prigogine I (1967) See literature list of Chapter I
Tyrell HJV (1961) Diffusion and heat flow in liquids. Butterworth's, London
Knudsen M (1950) See literature list of Chapter VII

Thermoosmosis

Pagliuca N, Mita DG, Gaeta FS (1983) J Membr Sci 14:31
Denbigh K, Raumann C (1952) Proc Roy Soc A 210:377 and 518
Alexander KF, Wirtz KZ (1950) Z Phys Chem (Leipzig) 195:165
Allen JF, Jones H (1938) Nature 141:243

Fountain Effect

Zemansky MW (1968) Heat and thermodynamics. 5th edn, McGraw-Hill, New York, p 512

Chapter X. Thermoelectric Phenomena

A. Introduction

The purpose of this chapter is the interpretation of the three major thermoelectric effects, viz., Seebeck's, Peltier's and Thomson's, in phenomenological terms, in analogy to the treatment of the coupling of mass with heat flow in Chapter IX. Thus, this chapter is to serve merely as an introduction and bridge to the substantial literature on thermoelectricity and its energetic aspects, e. g., solar photovoltaic cells or thermoelectric refrigerators.

B. Qualitative Discussion of the Coupling Effects

The basic concept for the Seebeck and Peltier effects, known since the first half of the 19th century, is the *thermocouple* or *thermoelement*, which is a loop of two wires made from different metals, A and B, or from two different semiconductors similarly connected. Seebeck discovered in 1821 that if the two junctions of the metals are at different temperatures, an electric potential difference develops between them. This discovery forms the basis of *thermoelectric power sources*. Although of different physical nature, the Seebeck effect is phenomenologically analogous to the thermomolecular pressure difference, which is due to the coupling of heat flow with *gas* flow, rather than with *electron* flow. Both the thermomechanical effect (Fig. IX-2) and the thermomolecular pressure difference occur in systems of the type $G|D|G'$ (G = gas; D = thermoselective diaphragm), and the Peltier and Seebeck effects (Figs. X-1 and X-2, respectively) occur in systems $A|B|A'$ (A = A' and B being two different electronic conductors).

The strength of the Seebeck effect depends on the nature of A and B. It is expressed by the *Seebeck coefficient*

$$\eta_{SB} \equiv \left[\frac{d(\varDelta \mathscr{E})}{d(\varDelta T)} \right]_{I=0}. \tag{X-1}$$

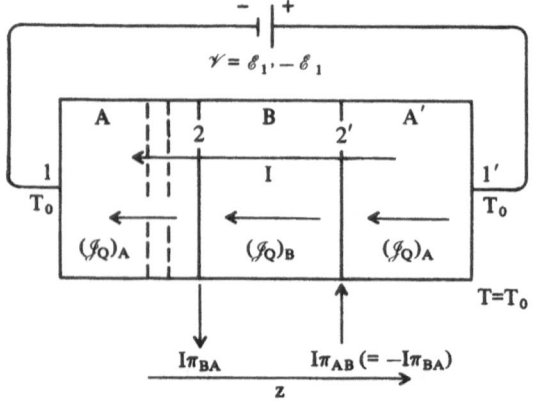

Fig. X-1. Peltier effect ($\Delta T = 0$). An electric current passes through an assembly of conductors A'BA. To keep the temperature constant, heat must be added at the BA' junction (*2'* heat sink), and withdrawn at the BA junction (*2* heat source). In this figure we have:

$$\mathscr{V} > 0, \ I < 0, \ \mathscr{J}_Q < 0, \ \pi_{AB} < 0,$$
$$\pi_{BA}(= -\pi_{AB}) > 0$$

This coefficient has been found to vary relatively little with the temperature of one junction, when the temperature of the other junction is held constant.[1]

The Peltier effect is schematically represented in Fig. X-1. Here the junctions of the thermocouple ABA' are placed into thermostats held at the same temperature. Peltier found 1834 that the passage of an electric current causes the uptake of heat, $(\mathscr{J}_Q)_P$, at one junction (BA' in Fig. X-1) and the rejection of heat at the other (AB). The amount of heat uptake is equal to the heat rejected if A = A' and the junctions are at the same temperature. These heat flows, which are both proportional to the electric current, I (A), are analogous to the heat flows into the refrigerator and from the heater in the thermomechanical-pressure experiment (Fig. IX-2).[2]

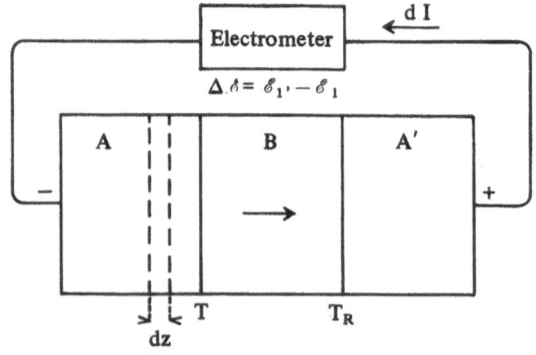

Fig. X-2. Seebeck effect ($\Delta T \equiv (T_R - T) \neq 0, \ T > T_R$). The junctions 2 and 2' are maintained at temperatures T and T_R, respectively. An electric potential difference, $\Delta \mathscr{E}$, appears. It is measured with a high-impedance voltmeter or electrometer, which draws the very small current, dI. When this thermoelement is used for the measurement of the temperature, T, T_R is sometimes held constant by immersion of junction 2' in an ice bath

---

1 It follows from Eqs. (IX-10) and (IX-24) that the analogous thermomolecular-pressure coefficient for ideal gases varies relatively little with the temperature when experiments are compared in which the molar volume of the gas, $\bar{V}_k$, is almost constant.

2 If the same thermocouple, ABA', were used as a thermoelectric power source (Seebeck effect), the temperature difference being imposed such that the generated electric current flows in the same direction as in the Peltier experiment, then the *hot junction* in the Seebeck experiment is the *heat-uptake junction* in the Peltier experiment. (Note that

The Peltier effect for a given thermocouple ABA is characterized by the *Peltier coefficient*, π_{AB}:

$$|\pi_{AB}| = |(\mathscr{J}_Q)_P/I|. \tag{X-2}$$

When the direction of the electric current is inversed, the direction of the heat flows is also inversed. The heat-uptake junction is the cooling element in a thermoelectric refrigerator.

While both the Seebeck and Peltier effects were observed in thermocouples, W. Thomson (later Lord Kelvin) postulated that even in a wire made entirely out of a single electric conductor, heat flow, $(\mathscr{J}_Q)_{Th}$, to or from the surroundings (*Thomson effect*, Fig. X-3) has to occur upon passage of an electric current, I, provided a temperature gradient is maintained along the conductor at the time of electric-current flow. This heat flow is superimposed on the heat flow from the conductor to the surroundings caused by the normal Joule heating effect, $I^2\mathscr{R}$, which takes place even in the absence of the Thomson effect, for which a temperature gradient is essential. The heat flow, $(\mathscr{J}_Q)_{Th}$, is proportional to the electric current.

For a given conductor, the Thomson effect is characterized by a parameter σ_{Th} (*Thomson heat*):

$$\sigma_{Th} \equiv (1/I)\,d(\mathscr{J}_Q)_{Th}/dT. \tag{X-3}$$

It is possible to make the existence of the Thomson effect plausible by drawing an analogy with the thermomechanical effect. Figure IX-2 describes this effect schematically. In order to *keep the temperature uniform* in an ideal gas

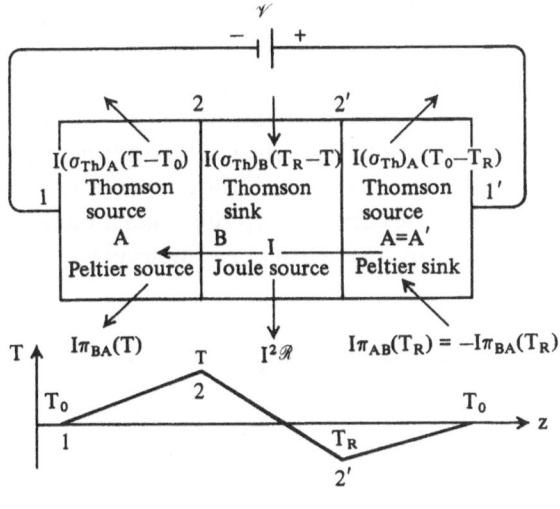

Fig. X-3. Thomson effect ($T_R < T_0 < T$). An electric current is passed through the assembly of conductors A'BA. The junctions 2 and 2' are held at temperatures T and T_R, respectively. The passage of current creates heat sources and sinks. Where the electric current goes from lower to higher temperature, heat is absorbed from the surroundings (Thomson sink). As in Fig. X-1, we have: I < 0; $\pi_{BA}(= -\pi_{AB}) > 0$

in Figs. X-1 and X-2, the conditions were chosen such that the respective electric currents flow in *opposite* directions.)

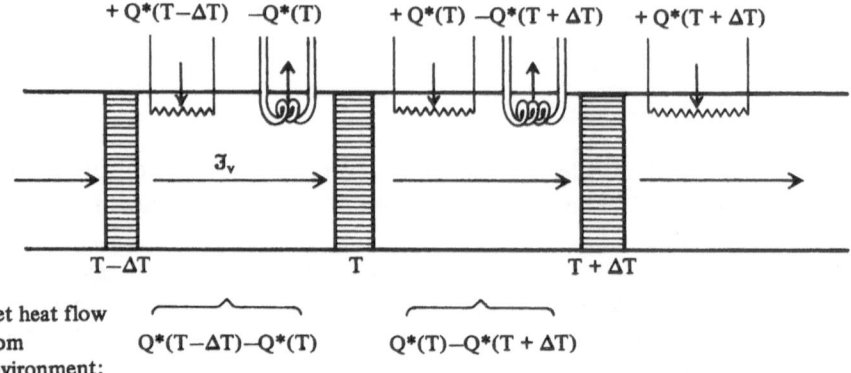

$$+ Q^*(T-\Delta T) \quad -Q^*(T) \qquad + Q^*(T) \quad -Q^*(T + \Delta T) \qquad + Q^*(T + \Delta T)$$

Net heat flow
from $Q^*(T-\Delta T)-Q^*(T)$ \qquad $Q^*(T)-Q^*(T + \Delta T)$
environment:

Fig. X-4. Thermomechanical analogy of the Thomson effect. Figure illustrates the lateral heat flows to the environment which are necessary to maintain a given temperature gradient along a thermoselective plug when a steady flow of an ideal gas through the plug takes place

permeating a thermoselective plug, it is necessary to refrigerate and heat the gas on the high- and low-pressure sides of the diaphragm, respectively. To draw an analogy with the Thomson effect, we consider a thermoselective porous plug across which *a temperature gradient is maintained* (Fig. X-4). Imagine that we subdivide the plug into slices of differential thickness, separated by compartments containing the gas. It is assumed that the temperature gradient occurs in these spaces, while the temperature change across the plug slices is made to be negligible by heating and refrigeration, respectively.

Comparing each plug slice to the thermoselective diaphragm in Fig. IX-2, we see that for each mol of gas permeating, the heat of transfer, Q^*, has to be removed on the left (high-pressure) face of the slice, and a numerically equal flow of heat has to be introduced on the right face of the slice. Because the heat of transfer varies with the temperature [Eq. (IX-24)], the heat flow into each gas compartment is not numerically equal to the heat outflow. Thus, a (lateral) net heat flow from the model to the surroundings has to take place if the temperature gradient is to be maintained, just as in an electric conductor (rather than a thermoselective porous plug) through which electrons (rather than gas molecules) flow.

C. Flow Equations and Relations Between Thermoelectric Phenomena

1. Formulation of the Flow Equations

The flow equations for the coupled heat flow, $\mathscr{J}_Q$, and the electric flow are of the type of Eqs. (V-2). The generalized driving forces, X_j, which are conjugated to these flows, are substituted in Eqs. (V-2) from Table III-1. Because of the different nature of conductors A ($=$ A$'$) and B, the conductance coefficients, L_{ij}, are different for the two conductors. Hence, there is a separate pair of coupled flow equations for each of the two conductors. The electric current, I, is equal in all conductors (Figs. X-1 to X-3), but the heat flow is not continuous; hence $(\mathscr{J}_Q)_A$ and $(\mathscr{J}_Q)_B$ are, in general, not equal:

$$I = (L_{II})_A\left(-\frac{1}{T}\frac{d\mathscr{E}}{dz}\right) + (L_{IQ})_A\left(-\frac{1}{T^2}\frac{dT}{dz}\right) \quad\left.\right\} \text{A} \tag{X-4}$$

$$(\mathscr{J}_Q)_A = (L_{QI})_A\left(-\frac{1}{T}\frac{d\mathscr{E}}{dz}\right) + (L_{QQ})_A\left(-\frac{1}{T^2}\frac{dT}{dz}\right) \tag{X-5}$$

$$I = (L_{II})_B\left(-\frac{1}{T}\frac{d\mathscr{E}}{dz}\right) + (L_{IQ})_B\left(-\frac{1}{T^2}\frac{dT}{dz}\right) \quad\left.\right\} \text{B}. \tag{X-6}$$

$$(\mathscr{J}_Q)_B = (L_{QI})_B\left(-\frac{1}{T}\frac{d\mathscr{E}}{dz}\right) + (L_{QQ})_B\left(-\frac{1}{T^2}\frac{dT}{dz}\right) \tag{X-7}$$

The reciprocity relations, $L_{IQ} = L_{QI}$, are satisfied in both pairs of flow equations.

2. Peltier Effect

Dividing Eqs. (X-5) or (X-7) by Eqs. (X-4) or (X-6), respectively, we obtain for isothermal conditions, e.g., those depicted in Fig. X-1, by setting $dT/dz = 0$:

$$\left[\frac{(\mathscr{J}_Q)_A}{I}\right]_T = \frac{(L_{QI})_A}{(L_{II})_A} \equiv \pi_A \tag{X-8}$$

$$\left[\frac{(\mathscr{J}_Q)_B}{I}\right]_T = \frac{(L_{QI})_B}{(L_{II})_B} \equiv \pi_B. \tag{X-9}$$

π_A and π_B are the heat flows (per amp) in conductors A and B, respectively, coupled to the electric current, I. If the coupling is positive, i.e., the electric current causes heat flow in the same direction, as in Fig. X-1, it follows that π_A and π_B are positive. These coupling parameters depend on the nature of the conductor and on the temperature. In the case shown in Fig. X-1, $\pi_B > \pi_A$.

Hence, the heat flow, $(\mathscr{J}_Q)_B$, arriving at junction 2 as a result of coupling to the electric current, I, through conductor B is larger than the heat flow removed from this junction as a flow coupled to the electric current I in A. Hence, to maintain isothermal conditions, the difference of these two heat flows, $I\pi_{BA}$, has to be withdrawn laterally to the environment as shown in Fig. X-1 (since I is negative and π_{BA} positive, the product $I\pi_{BA}$ is negative):

$$(\mathscr{J}_Q)_B - (\mathscr{J}_Q)_A = I(\pi_B - \pi_A) = I\pi_{BA} < 0 . \tag{X-10}$$

Reasoning in analogous manner about the heat-flow balance of junction 2′, we find that a numerically equal heat flow, $I\pi_{AB}$, has to be added laterally (from the environment) to this junction. The Peltier heat [Eq. (X-2)] is

$$\pi_{BA} \equiv \pi_B - \pi_A = -\pi_{AB} > 0 . \tag{X-11}$$

3. Seebeck Effect

It is of interest to calculate the electric potential difference, $(\Delta\mathscr{E})_{I=0}$, across a thermocouple, the junctions of which are at temperature T and the reference temperature, T_R, respectively (Fig. X-2). We first calculate the electric potential difference, $d\mathscr{E}_A$, across an element of the conductor A, subjected to a temperature difference, dT, when the electric current is negligible. We obtain from Eqs. (X-4) and (X-8)[3]:

$$(d\mathscr{E}_A)_{I=0} = -\frac{(L_{IQ})_A dT}{(L_{II})_A T} = -\frac{\pi_A}{T} dT \tag{X-12}$$

and similarly for conductor B:

$$(d\mathscr{E}_B)_{I=0} = -\frac{\pi_B}{T} dT . \tag{X-13}$$

We now sum all electric potential drops across the thermocouple, from left to right:

$$(\Delta\mathscr{E})_{I=0} = -\left[\int_{T_0}^{T}\frac{\pi_A}{T} dT + \int_{T}^{T_R}\frac{\pi_B}{T} dT + \int_{T_R}^{T_0}\frac{\pi_A}{T} dT\right]$$

$$= -\int_{T}^{T_R}\frac{\pi_B - \pi_A}{T} dT = \int_{T_R}^{T}\frac{\pi_{BA}}{T} dT = -\int_{T_R}^{T}\frac{\pi_{AB}}{T} dT . \tag{X-14}$$

---

3 The reciprocity relation $L_{IQ} = L_{QI}$ is used here.

The variation of this voltage of the thermocouple with the temperature T is found by differentiation with respect to T:

$$\left[\frac{d(\Delta \mathscr{E})}{dT}\right]_{I=0} = - \left[\frac{d(\Delta \mathscr{E})}{d(\Delta T)}\right]_{I=0} = \frac{\pi_{BA}}{T} = - \frac{\pi_{AB}}{T} > 0. \qquad (X-15)$$

Here, $-d(\Delta T) = -d(T_R - T)$, as seen from Fig. X-2, was substituted for dT. Introducing the Seebeck coefficient [Eq. (X-1)], we obtain

$$\eta_{SB} = \left[\frac{d(\Delta \mathscr{E})}{d(\Delta T)}\right]_{I=0} = \frac{\pi_{AB}}{T} = - \frac{\pi_{BA}}{T}. \qquad (X-16)$$

This equation, which establishes a quantitative relationship between the Seebeck and Peltier effects, is one of two basic Thomson[4] equations of thermo-electricity. The other Thomson equation, (X-23), is derived in Sect. 4.

4. Thomson Effect

To relate the Thomson coefficients to the voltage produced by a thermocouple ABA', the junctions of which are held at a (variable) temperature, T, and the reference temperature, T_R, respectively, (Fig. X-3) we first establish its energy balance in the steady state during the passage of an electric current, I. In order to maintain a constant temperature profile, it is necessary to add or subtract heat from the heat sinks and heat sources, respectively, shown in the figure. Because of energy conservation, the algebraic sum of all these heat flows and the electric-energy input, $I\mathscr{V}$, is zero[5]:

$$0 = -I\mathscr{V} - \mathscr{R}I^2 + I\pi_{BA}(T) + I\pi_{AB}(T_R) + I(\sigma_{Th})_B(T_R - T)$$
$$+ I(\sigma_{Th})_A(T - T_0) + I(\sigma_{Th})_A(T_0 - T_R). \qquad (X-17)$$

Combining similar terms, we obtain:

$$I\mathscr{V} = -\mathscr{R}I^2 + I\{\pi_{BA}(T) - \pi_{BA}(T_R) + [(\sigma_{Th})_A - (\sigma_{Th})_B](T - T_R)\}. \qquad (X-18)$$

To relate the phenomena shown in Fig. X-3 to the Seebeck effect, we apply Eq. (X-18) to the simpler situation shown in Fig. X-2 (i.e., extremely small

---

4 Since William Thomson was to become Lord Kelvin, the Thomson equations of thermo-electricity are also known as Kelvin's equations.

5 Note that in Fig. X-3, the current, I, flows from right to left and is therefore negative, by definition. $\mathscr{V}$ is the electric-potential difference between locations 1' and 1 (Fig. X-3).

electric current), we divide the equation by I and calculate the limiting value of the voltage for vanishing electric current:

$$\lim_{I \to 0} \mathscr{V} = (\Delta\mathscr{E})_{I=0} = \pi_{BA}(T) - \pi_{BA}(T_R) + [(\sigma_{Th})_A - (\sigma_{Th})_B](T - T_R). \quad \text{(X-19)}$$

Differentiating with respect to the variable temperature, T, while the reference temperature, T_R, is held constant, we obtain

$$\left[\frac{d(\Delta\mathscr{E})}{dT}\right]_{I=0} = -\left[\frac{d(\Delta\mathscr{E})}{d(\Delta T)}\right]_{I=0} = \frac{d\pi_{BA}}{dT} + (\sigma_{Th})_A - (\sigma_{Th})_B. \quad \text{(X-20)}$$

Another useful form of Eq. (X-20), viz., Eq. (X-23), is obtained by introducing a conclusion from Eq. (X-15) into it, as follows. Multiplying both sides of Eq. (X-15) by T, we obtain:

$$\pi_{BA} = T\left[\frac{d(\Delta\mathscr{E})}{dT}\right]_{I=0} = -T\left[\frac{d(\Delta\mathscr{E})}{d(\Delta T)}\right]_{I=0}. \quad \text{(X-21)}$$

We now differentiate Eq. (X-21) with respect to T:

$$\frac{d\pi_{BA}}{dT} = \left[\frac{d(\Delta\mathscr{E})}{dT}\right]_{I=0} = T\left[\frac{d^2(\Delta\mathscr{E})}{dT^2}\right]_{I=0}. \quad \text{(X-22)}$$

Substituting for $d(\Delta\mathscr{E})/dT$ in Eq. (X-20) from Eq. (X-22), we obtain

$$-T\left[\frac{d^2(\Delta\mathscr{E})}{dT^2}\right]_{I=0} = -T\left[\frac{d^2(\Delta\mathscr{E})}{d(\Delta T)^2}\right]_{I=0} = (\sigma_{Th})_A - (\sigma_{Th})_B. \quad \text{(X-23)}$$

The pair of equations, (X-16) and (X-23), are the basic *Thomson equations of thermoelectricity.*

Problems

X.1. In the temperature range $0-600°C$, the electromotive force, $\Delta\mathscr{E}$, of the thermocouple $Cu - Fe$ is:

$$\Delta\mathscr{E}_{Cu-Fe}(\mu V) = -13.4\Delta T - 0.01375(\Delta T)^2 + 0.000087(\Delta T)^3$$

with: $\Delta T = T_R - T;$ $T_R = 0°C;$ and $\Delta\mathscr{E} = \mathscr{E}_{1'} - \mathscr{E}_1.$

The notation is as in Fig. X-2, with A = Cu, B = Fe.

a) Calculate the Peltier heat transferred per hour at each junction for a current of I = -10^{-3} A, when the whole system is held at T = T_R = 300°C.
b) Same question for 100°C.
c) At what temperature between 100°C and 300°C is the Peltier heat zero?
d) Calculate the differences between the Thomson heats of iron and cooper, respectively:
(1) at 300°C, and (2) at 100°C.

X.2. The *thermoelectric power (TEP)* of a metal, M, is defined as its Seebeck coefficient with respect to pure lead (Pb). In terms of Fig. X-2, the couple would be represented as Pb − M − Pb', T_R = 0°C, and T is designated T_C.

The variation of TEP ≡ $d(-\Delta\mathscr{E})/dT_C$ with the temperature T_C is often represented by the linear expansion

$$\frac{d(\Delta\mathscr{E})}{d(\Delta T)} = \frac{d(-\Delta\mathscr{E})}{dT_C} = A + BT_C, \qquad (P.X.1)$$

where A and B are empirical constant depending on the nature of metal M and its mode of preparation.

The following constants A and B are found in the literature[6]:

M	Couple	A [µV/°C]	B [µV/°C^2]	Temperature range [°C]
Platinum	Pb − Pt	− 3.038	− 0.03248	− 200 → + 300
Silver (electrolytic)	Pb − Ag	+ 2.947	+ 0.006782	− 200 → + 100

a) Using the definitions and sign conventions of this chapter, show that the expression for the Seebeck coefficient corresponding to Eq. (P.X.1) is:

$$\eta_{SB} = A - B(\Delta T) = A + BT_C = (TEP). \qquad (P.X.2)$$

b) Show that the thermoelectric power of the couple $M_1 - M_2 - M_1'$, designated as $(TEP)_{M_1-M_2}$ equals $(TEP)_{Pb-M_1} - (TEP)_{Pb-M_2}$
Hint: Use Eqs. (X-16) and (X-11).
c) Express the electric potential difference, $\Delta\mathscr{E}$, between the junctions at T_R and T, respectively, of the platinum-silver couple as a function of the temperature.

---

6 Handbook of Chemistry and Physics, 34th edn, Chemical Rubber Publishing Co, Cleveland Ohio (1952) pp 2218 – 19.

Selected Literature

Texts on Thermoelectricity

MacDonald DKC (1967) Thermoelectricity – an introduction to the principles. Wiley,
  New York
Barnard RD (1972) Thermoelectricity in metal and alloys. Halsted Press, New York
Zemansky MW (1968) Ch 13: see literature list of Chapter I
Heikes RR, Ure PW Jr (1961) Thermoelectricity: science and engineering. Interscience,
  New York
Joffe AF (1957) Semiconductors, thermoelements and thermoelectric cooling. Infosearch
  Ltd, London

Some Other Literature Relevant to Chapter X

Russel CR (1967) Elements of energy conversion. Pergamon Press, London
Munster A (1966) Thermodynamique des processus irréversibles. Presses Universitaires de
  France, Paris
Wright DA (1965) In: Spring KH (ed) Direct generation of electricity. Academic Press,
  New York
Thomson WM (1854) Proc Roy Soc Edinburgh 3:225

List of Symbols

Symbol	Units	
a	cm^2	surface
A, A′, B, C		components
$\mathscr{A}$	$W\ s\ mol^{-1}$ (*)	affinity
c	$mol\ cm^{-3}$	concentration in gas or solution
$\bar{c}$	$mol\ (cm^3\ membrane)^{-1}$	concentration in separator (membane)
D	$cm^2\ s^{-1}$	diffusion coefficient in gas or solution
D^a	$cm^2\ s^{-1}$	self-diffusion coefficient
$\bar{D}$	$cm^2\ s^{-1}$	diffusion coefficient in separator (membrane)
$\bar{e}*$	$W\ s\ mol^{-1}$	energy carried by one mol of component
E	$W\ s$	total energy
$\bar{E}$	$W\ s\ mol^{-1}$	partial molal energy
$\mathscr{E}$	V	electric potential
f_{ij}	$W\ s^2\ cm^{-2}\ mol^{-1}$	friction coefficient of species i with j
f_M	dimensionless	geometric coefficient correcting for non-perpendicular incidence of gas molecules on wall
f_G	dimensionless	coefficient correcting for reflection of gas molecules incident on porous separator
F_{Helm}	$W\ s$	Helmholtz free energy
$_{\Delta}F$	$W\ s$ per unit of transported species	(Rayleigh) generalized driving force (used for isothermal processes)

(*) 1 W s = 1 J.

Symbol	Units	
F	W s cm^{-1} per unit of transported species	*differential* (Rayleigh) driving force
$\mathscr{F}$	C Eq^{-1}	Faraday's constant (9.65×10^5)
G	W s	(Gibbs) free energy (enthalpie libre)
g	dimensionless	rational osmotic coefficient
H	W s	total enthalpy
H$_u$	W s	(tabulated) enthalpy of a simple system at rest
$\bar{H}$	W s mol^{-1}	partial molal enthalpy
i	A cm^{-2}	electric current density
I	A	electric current
j$_{chem}$	mol cm^{-3} s^{-1}	rate of progression of isomerization reaction in unit volume
J$_x$	units of species x cm^{-2} s^{-1}	flux of species x
$\mathscr{J}_{chem}$	mol s^{-1}	rate of progression of chemical reaction
$\mathscr{J}_x$	units of species x s^{-1}	flow of species x
k	s^{-1}	rate constant for first-order reaction
k	W s K^{-1} molecule^{-1}	Boltzmann's constant (1.38×10^{-23})
K$_{eq}$	see Eq. (VI-15)	chemical-reaction equilibrium constant
K$_{JT}$	K cm^3 W^{-1} s^{-1} = MPa^{-1} K	Joule-Thomson coefficient
K$_K$	K^{-1} cm^{-3} W s = MPa K^{-1}	thermomolecular pressure coefficient (Knudsen coefficient)
K	W s cm^{-3} K$^{-0.5}$	coefficient defined in Eq. (IX-18)
L	(units of transported species)^2 cm^{-1} W^{-1} s^{-2} (K)*	(Onsager's) conductance coefficient [Eq. (V-2)]
L$_p$	cm^4 s^{-2} W^{-1} = cm s^{-1} MPa^{-1}	hydraulic permeability

* When Rayleigh forces, F, are used [Eq. (V-1)], delete the term (K).

Symbol	Units	
m_e	$cm^2\ V^{-1}\ s^{-1}$	electric mobility
$m_{d,k}$	$mol\ k\ cm^2\ W^{-1}\ s^{-2}$	diffusive mobility of k
$\mathbf{m}$	$g\ molecule^{-1}$	molecular mass
n_k	mol	number of mol of k
$\dot{N}$	$molecule\ s^{-1}\ cm^{-2}$	number of molecular impacts per unit area and s
p	$W\ s\ cm^{-3} = MPa*$	pressure
$\mathscr{P}_I$	$W^{-1}\ s^{-2}\ cm^5 = cm^2\ s^{-1}$ MPa^{-1}	specific *hydraulic* permeability
$\bar{P}$	$cm^2\ s^{-1}$	specific permeability to *solute*
Q	$W\ s**$	heat input
$Q*$	$W\ s\ mol^{-1}$	heat of transfer
R	$W\ s\ mol^{-1}\ K^{-1}$	universal gas constant (8.314)
$\mathscr{R}$	Ω	electric resistance
$\mathscr{R}_H$	$W\ s^2\ cm^{-4}$	hydraulic resistance
R	$W\ cm\ s^2$ (units of transported species)^2	(Rayleigh) resistance coefficient [Eq. (IV-6)]
$\mathbf{R}$	dimensionless	rejection coefficient of membrane
$\dot{s}$	$W\ K^{-1}\ cm^{-3}$	rate of entropy production per unit volume
S	$W\ s\ K^{-1}$	entropy
$\bar{S}$	$W\ s\ K^{-1}\ mol^{-1}$	partial molal entropy
$\dot{S}$	$W\ K^{-1}$	rate of entropy production
t	s	time
T	K	(absolute) temperature
u	$cm\ s^{-1}$	velocity
U	$W\ s$	internal energy
γ	V	voltage (electric-potential difference)
V	cm^3	volume
$\bar{V}$	$cm^3\ mol^{-1}$	partial molal volume
W	$W\ s$	work output

* 1 MPa = 9.87 atm
** 1 W s = 0.239 cal

Symbol	Units	
$_\Delta X$	W s K^{-1} per unit of transported species	generalized driving force [Eq. (III-1)]
X	W s K^{-1} cm^{-1} per unit of transported species	*differential* generalized driving force
y	dimensionless	molar fraction
y_a	dimensionless	semipermeable surface fraction of membrane
y_b	dimensionless	non-selective surface fraction of membrane
z	cm	length
Z	equivalent mol^{-1}	ionic valence ($Z_{Na} = 1$, $Z_{SO_4} = -2$)
α_1	W^{-1} s^{-2} mol^2	ratio of isomerization reaction rate to affinity [Eq. (VI-29)]
β	cm^3 A^{-1} s^{-1}	electroosmotic coefficient
β''	cm W s^2 mol^{-2}	mutual friction coefficient of two gases
Δ		difference operator (right minus left)
$_{eq}\Delta$		denotes deviation from equilibrium [Eqs. (VI-20), (VI-27)]
ε	dimensionless	porosity (volume fraction of pore space)
ζ_{ij}	cm W s^2 mol^{-2}	mutual (symmetrical) friction coefficient between 1 mol of i and 1 mol of j
η_{SB}	V K^{-1}	Seebeck coefficient
$\varkappa'$	A V^{-1} cm^{-2} = Ω^{-1} cm^{-2}	conductance per unit surface
Λ	W s	exergy
$\dot{\Lambda}$	W	rate of exergy change
μ_i^c	W s mol^{-1}	chemical potential of component i (usually in a multicomponent system) *at uniform temperature and pressure*
μ^0	W s mol^{-1}	standard chemical potential
μ_i	W s mol^{-1}	chemical potential of i

Symbol	Units	
$\tilde{\mu}$	W s mol^{-1}	(Gibbs) general (total) potential
ν_i	mol produced (reaction)$^{-1}$	stoichiometric coefficient
ν	particles molecule^{-1}	number of particles produced by dissociation of one molecule
$d\xi/dt$	mol s^{-1}	degree of advancement of chemical reaction [Eq. (VI-5)]
Π	W s cm^{-3} = MPa	osmotic pressure
π_{AB}	W A^{-1}	Peltier heat
π_A, π_B	W A^{-1} = V	heat of transfer of electric charges flowing in conductors A and B, respectively
ρ_e	Ω cm	(electric) resistivity
σ	dimensionless	reflection coefficient of membrane
σ_{Th}	W s C^{-1} K^{-1}	Thomson heat
Φ_Λ		exergetic efficiency
ω	mol W^{-1} cm s^{-2}	solute permeability
Ω	\$ (or other currency) cm^{-3}	cost per unit volume of product

Superscripts

0	standard state
c	at given T, p the parameter having this superscript depends only on the concentration
1a	atmospheric pressure
sol	solution
c_s	at solution concentration c_s

Subscripts

A, B ⎱ **1, 2** ⎰	these subscripts designate different states
i, j, k ⎫ α, β ⎬ A, B ⎭	these subscripts designate different components

Symbol	Units
γ	mobile species in membrane or solution
Δ	when placed in lower left corner of a symbol, this subscript indicates that the parameter is defined for a difference rather than a derivative
0	reservoir
i	when placed next to an operator, this subscript denotes *internal* changes in a system
eq	equilibrium
K	Knudsen conditions
m	membrane
s	solute (or counterion in membrane)
Q	relating to heat flow
u	useful
u	simple system
w	water
JT	Joule-Thomson conditions
SB	Seebeck-effect parameter
Th	Thomson-effect parameter

Overbars

Overbars on extensive parameters, e.g., $\bar{H}$, designate partial molar quantities. Overbars on intensive parameters designate properties in membranes or diaphragms, e.g., $\bar{c}_i =$ mols i/total volume of diaphragm, pore volume included.

Sign Conventions

1. Positive direction is from left to right.
2. Fluxes from left to right are counted positive.
3. The operator, Δ, for finite differences refers to the value on the right minus the value on the left, as does conventionally the differential operator, d.

Symbol	Units

Driving forces are of the general form $(-d\tilde{\mu}/dz)$. Thus, *positive* values of the driving force, $(-d\tilde{\mu}/dz) > 0$, lead to *positive fluxes*. For example, Ohm's law is written as

$$i = (1/\rho) \underbrace{(-dE/dz)}_{\text{Driving force}}$$

and Fick's law

$$J_i = D_i \underbrace{(-dc_i/dz)}_{\text{Driving force}}$$

Appendix I. Carnot's Engine

A. Introduction

This engine operates under reversible conditions in a bithermal cycle, i. e., at two different stages of its 4-step cycle, it is in contact with reservoirs of two different temperatures, T'' and T', respectively ($T'' > T'$). The engine is simply a cylinder with a movable, frictionless piston, containing one mol of an ideal gas. Although this engine is not used in practice, the calculation of its efficiency is a valuable exercise, because Carnot has demonstrated that *no bithermal engine working between T'' and T' can have a higher efficiency* (defined as the ratio between the work flow produced and the heat input flow from the high-temperature reservoir into the engine), nor can *any other engine* working under *reversible* conditions between T'' and T' have a lower efficiency.

Moreover, the calculation of the efficiency illustrated that in a complete cycle $\mathscr{J}_{Q,rev}''/T'' = -\mathscr{J}_{Q,rev}'/T'$, where $\mathscr{J}''$ and $\mathscr{J}'$ are the heat flows between engine and reservoirs T'' and T', respectively ($\mathscr{J}''$ is counted positive, because it enters the engine, while $\mathscr{J}'$, which leaves the engine, is counted negative). In other words, the parameter $Q_{rev}/T \equiv \mathscr{J}_{Q,rev}''/T''$ is *conserved* in the global system consisting of the Carnot engine and the two reservoirs. This finding drew attention to the importance of this parameter, named *entropy* by R. Clausius, which was shown to be conserved not only in the Carnot system, but in all global systems in which reversible changes occur.

To calculate the efficiency of the reversible engine, and to demonstrate the entropy conservation in this cycle, we first prove an essential elementary theorem about the reversible expansion or compression of ideal gases.

B. Preamble to Efficiency Calculation

Theorem. When an ideal gas expands adiabatically and reversibly from T'' to T', the ratio between initial and final volume is independent of the initial volume.

Proof. Consider one mol of an ideal gas (temperature T'', molal volume $\bar{v}_i''$) expanding adiabatically to temperature T', as shown schematically in Fig. A-1.

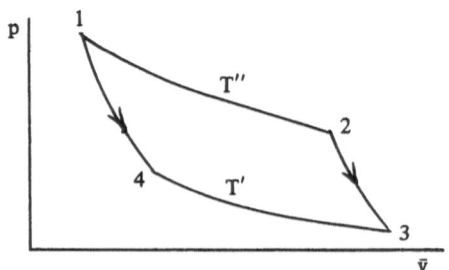

Fig. A-1. Two reversible adiabatic expansions between temperatures T″ and T′ (1 mol of an ideal gas)

The molal volume has increased to $\bar{v}_4''$. The law relating volumes to temperatures in reversible, adiabatic expansions of ideal gases is[1]

$$T\bar{v}^{\gamma-1} = \text{const} ,\qquad\qquad\text{(A-I-1)}$$

where γ is the specific-heat ratio c_p/c_v. Hence

$$T''\bar{v}_1^{\gamma-1} = T'\bar{v}_4^{\gamma-1} .\qquad\qquad\text{(A-I-2)}$$

Similarly, if one mol of ideal gas of the same initial temperature, T'', at a different pressure and at volume $\bar{v}_2$ (Fig. A-1) expands until its temperature reaches T':

$$T''\bar{v}_2^{\gamma-1} = T'\bar{v}_3^{\gamma-1} ,\qquad\qquad\text{(A-I-3)}$$

where $\bar{v}_3$ is the final molal volume in the second expansion process. Expressing the temperature ratio T'/T'' in terms of the volume ratio, we obtain from Eq. (A-I-2) and then from (A-I-3):

$$T'/T'' = (\bar{v}_2/\bar{v}_3)^{\gamma-1} = (\bar{v}_1/\bar{v}_4)^{\gamma-1} .\qquad\qquad\text{(A-I-4)}$$

If follows that

$$\bar{v}_2/\bar{v}_3 = \bar{v}_1/\bar{v}_4 .\qquad\qquad\qquad\text{Q.E.D.}\qquad\qquad\text{(A-I-5)}$$

Also, it follows from Eq. (A-I-5) that

$$\bar{v}_2/\bar{v}_1 = \bar{v}_3/\bar{v}_4 .\qquad\qquad\text{(A-I-6)}$$

---

1 See, for instance, Jones and Hawkins, 1965, pp 168 – 169.

C. Energy Balance and Efficiency of the Carnot Cycle

The cycle consists of the following *reversible* steps (Fig. A-2):
a) One mol of the ideal gas at T'' expands *isothermally* from volume $\bar{v}_1$ to $\bar{v}_2(T'')$.
b) This is followed by an adiabatic expansion to $\bar{v}_3(T')$.
c) The gas is isothermally compressed to volume $\bar{v}_4(T')$.
d) This is followed by adiabatic compression to the original volume $\bar{v}_1(T'')$.

In step (a), the heat flow, Q'', from the reservoir at T'' into the engine[2] is equal to the work, $W_{1\to2}$, produced by the gas as it expands reversibly, since the internal energy of an ideal gas at constant temperature is independent of pressure or volume, i.e., ΔE in Eq. (I-6) is zero:

$$Q'' = Q_{1\to2} = W_{1\to2} = \int_1^2 p\,d\bar{v} = \int_1^2 (RT''/\bar{v})\,d\bar{v} = RT''\ln(\bar{v}_2/\bar{v}_1) > 0\,, \quad \text{(A-I-7)}$$

where R is the universal gas constant.

In step (b) the gas expands *adiabatically*. Hence, by definition, $Q_{2\to3} = 0$. Therefore, from Eq. (I-6):

$$W_{2\to3} = -\Delta E_{2\to3} = U'' - U' > 0\,. \quad \text{(A-I-8)}$$

In step (c) the gas is compressed isothermally at T' from $\bar{v}_3$ to $\bar{v}_4$, i.e., work is *consumed* by the engine. In analogy to Eq. (A-I-7), this work is

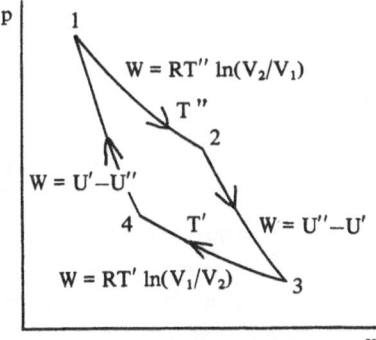

Fig. A-2. Carnot cycle

---

2 As in the main body of this text, heat flows *into* the system (in this case, the engine) are counted positive and so is work *produced* by the engine.

$$Q' = Q_{3\to4} = W_{3\to4} = \int_3^4 (RT'/\bar{v})\, d\bar{v} = RT' \ln(\bar{v}_4/\bar{v}_3) = -RT' \ln(\bar{v}_3/\bar{v}_4) < 0 .$$

$$\text{(A-I-9)}$$

The volume, $\bar{v}_4$, at which this compression stops, is on purpose selected in such manner that one additional adiabatic compression [*step (d)*] brings the gas back to its original state $(T', \bar{v}_1)$. In this final step of the cycle, $Q_{4\to1} = 0$ by definition. Hence, from Eq. (I-6):

$$W_{4\to1} = -\Delta E_{4\to1} = U' - U'' < 0 . \qquad\qquad \text{(A-I-10)}$$

Summing over all four steps of the cycle we find that

$$\Sigma\, W = W_{1\to2} + W_{2\to3} + W_{3\to4} + W_{4\to1} = RT'' \ln(\bar{v}_2/\bar{v}_1) - RT' \ln(\bar{v}_3/\bar{v}_4) .$$

$$\text{(A-I-11)}$$

Since the two volume ratios are equal [Eq. (A-I-6)], it follows that

$$\Sigma\, W = [R \ln(\bar{v}_2/\bar{v}_1)]\,(T'' - T') > 0 . \qquad\qquad \text{(A-I-12)}$$

We now apply Eq. (I-6) to the full cycle. Since the gas returned to its original state, $\Delta E_{\text{full cycle}} = 0$. Hence,

$$\Sigma\, Q = [\Sigma\, W] = [R \ln(\bar{v}_2/\bar{v}_1)]\,(T'' - T') . \qquad\qquad \text{(A-I-13)}$$

The efficiency of the Carnot process, η_C, is obtained from Eqs. (A-I-7) and (A-I-12):

$$\eta_C \equiv \Sigma\, W/Q'' = (T'' - T')/T'' = 1 - (T'/T'') . \qquad\qquad \text{(A-I-14)}$$

Comparing Eqs. (A-I-7) and (A-I-9), we see that

$$Q''/T'' = -Q'/T' , \qquad\qquad \text{(A-I-15)}$$

because, according to Eq. (A-I-6), the volume ratios $\bar{v}_2/\bar{v}_1$ and $\bar{v}_3/\bar{v}_4$ are equal.

Also, comparison of Eqs. (A-I-7) and (A-I-9) shows that, since $T'' > T'$, the heat inflow in step (a) is numerically larger than the heat outflow from the engine in step (c). The work output in step (a) is numerically larger than the work input necessary in step (c). Thus, there is a net absorption of heat and a net output of work in a full cycle; this is, after all, the purpose of designing a thermal engine:

$$\Sigma\, Q = \Sigma\, W > 0 . \qquad\qquad \text{(A-I-16)}$$

However, it follows from Eq. (A-I-15) that

$$Q''/T'' - Q'/T' = \sum (Q_{rev}/T) = 0 .\tag{A-I-17}$$

While the entropy of the *gas* changes in steps (a) and (c), these two changes cancel each other. Moreover, since, by definition, there is no entropy change in the reversible, adiabatic steps (b) and (d) ($Q_{rev} = 0$), no entropy is produced in a full cycle comprising all four reversible steps. As for the *reservoirs,* the loss of entropy from the upper reservoir, $-Q''/T''$, is counterbalanced by the numerically equal entropy gain of Q'/T' of the low-temperature reservoir. Thus, the operation of a Carnot engine causes no entropy change in the global system consisting of the engine and the two reservoirs.

D. Reverse Carnot Engine (Ideal "Heat Pump")

If the Carnot cycle is reversed, the heat and work outputs become inputs. In other words, work, $- \sum W$, is consumed in each cycle, while the low-temperature reservoir loses heat, $|Q'|$, and the high-temperature reservoir gains the larger amount of heat $Q'' = |Q'| + W$. Therefore, the ideal heat pump acts like a refrigerator, subtracting heat from the low-temperature reservoir.

The coefficient of performance of any refrigerator is defined as

$$\beta \equiv |Q'|/|\sum W| .\tag{A-I-18}$$

Substituting for $\sum W$ from Eq. (A-I-14), and then for Q' from Eq. (A-I-15), we obtain:

$$\beta_C = T'/(T'' - T') .\tag{A-I-19}$$

Carnot's reasoning led to a generalization analogous to his conclusion about the efficiency of the Carnot engine, viz., that no refrigerator working between T'' and T' can have a higher coefficient of performance, and that no other refrigerator operating under *reversible* conditions between these temperatures can have a lower coefficient of performance.

Appendix II. Dependence of Chemical Potential on Concentration [proof of Eq. (III-8)]

When a differential change is made in a simple multicomponent system (e. g., a mixture of gases in a cylinder with movable piston) of volume V, containing n_1, $n_2 \ldots n_k$ mols of components $1, 2 \ldots k$, the change of its free energy is expressed by the *Gibbs-Duhem equation*[1]

$$dG = V dp - S dT + \sum_{i}^{k} \mu_i^c dn_i, \tag{A-II-1}$$

where p, T and S are the pressure, absolute temperature and entropy of the system, respectively. Hence the change of the free energy with pressure at constant composition is

$$(\partial G/\partial p)_{T, n_1, n_2 \ldots n_k} = V \quad \text{(Poynting's equation)} . \tag{A-II-2}$$

Consider now the compression of the multicomponent system and calculate the influence of pressure on the chemical potential, μ_i, of component i, which is the partial molar free energy of i:

$$(\partial \mu_i/\partial p)_{T, n_i} = \partial/\partial p \left[(\partial G/\partial n_i)_{p, T, n_i'} \right] = \partial/\partial n_i \left[(\partial G/\partial p)_{T, n_i} \right]_{T, p, n_i'} . \tag{A-II-3}$$

Here, subscripts n_i and n_i' stand for constant composition (all n's constant)[2] and keeping all n's *except* n_i constant, respectively, and use was made of the fact that the order of the double differentiation of μ_i may be reversed without affecting the final result.

Substituting Eq.(A-I-2) in (A-II-3), we obtain

$$(\partial \mu_i/\partial p)_{T, n_i} = (\partial V/\partial n_i)_{p, T, n_i'} \equiv \bar{v}_i, \tag{A-II-4}$$

where $\bar{v}_i$ is the partial volume of i at temperature T and total pressure p (see also Problem I.2.b).

---

1 See, for instance, Problem III.1 of this text, and Katchalsky and Curran 1965, Eq. (5-46).
2 Subscript n_i is a shortened form of subscript $n_1, n_2 \ldots n_k$ in Eq. (A-II-2).

If the multicomponent mixture is composed of n mols *of ideal gases*, then the *partial molar* volumes[3] of *all* components are equal to the *molar volume* of any pure component at the total pressure, p, i.e., RT/p. Hence

$$(\partial \mu_i / \partial p)_{T, n_i} = \bar{v}_i = V/n = RT/p \, . \tag{A-II-5}$$

Dividing both sides of Eq. (A-II-5) by the mol fraction of i, $X_i \equiv n_i/n$, and remembering that the *partial pressure*, p_i, of the ideal gas i is defined[4] as

$$p_i \equiv X_i p \tag{A-II-6}$$

we obtain

$$(\partial \mu_i / \partial p_i)_{T, n_i} = RT/p_i \, . \tag{A-II-7}$$

Integrating, at constant temperature and composition of the gas mixture, from a given standard state (say, $T = 25\,°C$, $p = 1$ atm) at which the chemical potential of pure i is defined as $(\mu_i^0)'$, to any value of p_i, we obtain

$$\mu_{i,g} = (\mu_i^0)' + RT \ln(p_i) = (\mu_i^0)' + RT \ln(X_i p)$$
$$= (\mu_i^0)' + RT \ln[(n_i/n)(RT/V)] \, , \tag{A-II-8}$$

where subscript g indicates the gas mixture.

The molar concentration of i in the gas mixture is defined as

$$c_{i,g} \equiv n_i/V \, . \tag{A-II-9}$$

Introducing this concentration into Eq. (A-II-8), we obtain

$$\mu_{i,g} = \mu_{i,g}^0 + RT \ln(c_{i,g}) \, , \qquad . \tag{A-II-10}$$

where a new parameter has been defined:

$$\mu_{i,g}^0 \equiv (\mu_{i,g}^0)' + RT \ln(RT/n) \, . \tag{A-II-11}$$

---

3 The *partial molar* volume of a component, i, is defined as the change of the total volume of a large amount of the multicomponent system (at p, T) when one additional mol of i is added at pressure p and temperature T. The amount of the mixture must be large enough so that the change of all concentrations caused by this addition is negligible.

4 See, for instance, Jones and Hawkins 1963, p 392. Note, however, that the term partial volume used in that reference is not identical with the partial *molar* volume, which is used here.

In principle, $\mu_{i,g}^0$, which represents the chemical potential of i in the gas mixture, when the concentration of i is unity, depends on the temperature and on the partial pressures of the other gases. However, by definition, in *perfect gas mixtures*, each component behaves as though is existed alone at the temperature and pressure of the mixture; the dependence of the chemical potential of component i on its own concentration is given by Eq. (A-II-10), in which $\mu_{i,g}^0$ depends only on the temperature and not on the pressure. Hence this equation represents the concentration dependence of the chemical potential of i at a given temperature, irrespective of the presence of other ideal gases.

$$\mu_{i,g}^c = \mu_{i,g}^0 + RT \ln(c_{i,g}) .$$ (A-II-12)[5]

To demonstrate the validity of this type of relationship for ideal *solutions*, we consider a 2-phase system consisting of the gas mixture and a liquid phase, which is a solution of component i, in equilibrium with the gas phase. When pressure and temperature are uniform, the distribution of component i between the phases is described by

$$(c_{i,g}/c_{i,l})_T = K_i ,$$ (A-II-13)

where $c_{i,g}$ and $c_{i,l}$ are the concentratios of component i in the gas and solution phases, respectively, and K_i is the distribution coefficient between gas and liquid phase, which is often constant for *dilute* solutions in equilibrium with perfect gas mixtures (Henry's law).

At equilibrium, the pressures and temperatures of the two phases are equal, and the chemical potentials of any component are uniform (Gibbs' equilibrium criterion):[6]

$$\mu_{i,l}^c = \mu_{i,g}^c .$$ (A-II-14)

Substituting for the chemical potential of i in the gaseous phase from Eq. (A-II-12), we obtain

$$\mu_{i,l}^c = \mu_{i,g}^0 + RT \ln(c_{i,g}) = \mu_{i,g}^0 + RT \ln(K_i) + RT \ln(c_{i,l})$$
$$= \mu_{i,l}^0 + RT \ln(c_{i,l}) ,$$ (A-II-15)

where the standard potential of i in solution is defined as

$$\mu_{i,l}^0 = \mu_{i,g}^0 + RT \ln(K_i) .$$ (A-II-16)

---

5 The superscript c was added to indicate that only changes in systems at uniform pressure and temperature are considered in the following discussions in this section.
6 See Problem II.1.

In principle, the chemical potential of a component in liquid solution varies with the pressure, but since the partial molar volume of a component in solution is generally much smaller than its molar volume in the gaseous state, this dependence [Eq. (A-II-4)] is often negligible.

It is seen from Eq. (A-II-15) that the assumptions about the perfect nature of the gas mixture and validity of Henry's law led to Eq. (III-8).

Q.E.D.

Appendix III. Answers to Problems

I.1.

a) Imagine a body of volume V and density ρ_s resting at the bottom of a thermally insulated container filled with liquid of density $\rho_l (\rho_l < \rho_s)$ to height h. Lift the body out of the liquid. Because of energy conservation in the system consisting of the body and the liquid, the work invested, $-W$, is equal to the change of gravitational energy, viz., the gain of potential energy of the body, $V\rho_s gh$, minus the loss of potential energy, $V\rho_l gh$, of an equal volume of liquid, which manifests itself in a drop of the water level upon extraction of the body from the liquid:

$$\Delta E = V\rho_s gh - V\rho_l gh = -W = Fh = V\rho_{app}gh,$$

where F is the force opposing the lifting of the immersed body, and ρ_{app} is the *apparent* density of the body when immersed in the liquid. Dividing this equation by h, we obtain the apparent weight of the body, $V\rho_{app}$:

$$V\rho_{app}g = V\rho_s g - V\rho_l g = \text{weight of body in vacuo} - \text{weight of displaced liquid}$$

Q.E.D.

b) 1. $\Delta E = 0$
   2. $\Delta T = 2.87 \times 10^{-6}$ K
   3. $\Delta S = 4.57 \times 10^{-6}$ W s (K)$^{-1}$ = 1.091 $\times 10^{-6}$ cal (K)$^{-1}$

I.2. For these proofs, it is useful a) to keep in mind that the chemical potential of component i is the partial molar Gibbs free energy and b) to use the Gibbs equation [see, for instance, Hatsopoulos and Keenan 1965, p 272].

I.3.

$$\Delta E = \Delta U = -100 \text{ cal mol}^{-1}$$
$$\Delta H = -140 \text{ cal mol}^{-1}$$
$$\Delta S = 0.0339 \text{ cal mol}^{-1} \text{ (K)}^{-1}$$

I.4.

a) $-W = KTl_0$.

b) Express the (reversible) heat transfer, Q_{rev}, to the surroundings in terms of $(\partial f/\partial T)_l$. In the equation of state, $f(T,l)$, differentiate f with respect to T, keeping in mind that *both* l and l_0 depend on T. Substitute the derivative in Q_{rev}.

c) $\Delta E = (5/2) K T^2 l_0 \alpha_0$.

d) $\Delta E = 0$. Hence $- W_{rev} = Q_{rev}$.

I.5.

a) $\Delta H_w = - 1346$ cal mol^{-1}

   $\Delta S_w = - 4.92$ cal mol^{-1} (K)$^{-1}$

   $\Delta G_w = - 52$ cal mol^{-1}

b) $\Delta H_{res} = 1346$ cal mol^{-1}

   $\Delta S_{res} = 5.12$ cal mol^{-1} (K)$^{-1}$

   $\Delta G_{res} = 0$.

c) For the global system: $\Delta H = 0$, $\Delta S = 0.21$ cal mol^{-1} (K)$^{-1}$, $\Delta G = - 52$ cal mol^{-1}.

   $\Delta \Lambda = \Delta G$ (taking 263 K as the temperature of the reference reservoir).

II.1. If this were not true, then the interchange of a small amount of any component between the phases would cause a change of free energy of the system, which, in turn, could be utilized for the production of useful work. But a system at equilibrium cannot produce such work (Postulate 1, Chap. I).

II.2.

a) $W_u = - 6683$ W s mol^{-1}

b) $\bar{\Lambda}'' = 3083$ W s mol^{-1}

   $\bar{\Lambda}' = 1665$ W s mol^{-1}

c) $W_{u, max} = 1418$ W s mol$^{-1} = \bar{\Lambda}'' - \bar{\Lambda}'$

II.3.

a) Split the entropy and exergy changes of system A into two parts, namely (1) the changes due to a (hypothetical) reversible transfer of heat from A to the reservoir and (2) the changes due to irreversible processes *within* A. For the latter, the Maxwell-Gouy-Stodola relationship [Eq. (II-33)] is valid.

b) For the global system, there are two sources of exergy loss, namely the loss due to the irreversible process within A plus the exergy loss due to the irreversibility of the heat transfer [Eq. (II-29)]:

$$\Delta \Lambda_{global} = - T_0 \Delta_i S_A + Q [1 - (T_0/T_A)] .$$

II.4.

a) From Eq. (II-14): $\mathscr{A}_W'' = 1000 (\bar{H}_i' - \bar{H}_i'')$

b) $\bar{S}_i' = \bar{S}_i''$.

II.5. With respect to reservoir a): $\bar{\Lambda} = 17122$ W s mol$^{-1} = 4090$ cal mol^{-1}.

With respect to reservoir b): $\bar{\Lambda} = 21475$ W s mol$^{-1} = 5130$ cal mol^{-1}.

III.1.

$(\partial \mu_A / \partial z)_{p,T} = RT d \ln(c_A) = RT dc_A / c_A$

$(\partial \mu_B / \partial z)_{p,T} = RT dc_B / c_B$

p is uniform; so is $c \equiv c_A + c_B$. Hence $dc_A = -dc_B$.

Therefore $[(\partial \mu_A / \partial z)/(\partial \mu_B / \partial z)]_{p,T}[=(\partial \mu_A / \partial \mu_B)_{p,T}] = -c_B / c_A$ Q.E.D.

III.2.

a) $D_{KCl} = 0.74 \times 10^{-5} \, cm^2 \, s^{-1}$

Resistance factor of disc = 2.57

$\dot{s}_d = 0.909 \times 10^{-7} \, W \, (K)^{-1} \, cm^{-3}$

$\dot{\lambda} = 2.71 \times 10^{-5} \, W \, cm^{-3}$

b) $\Delta \mathscr{E} = 2.58 \times 10^{-2} \, V$

c) $\dot{s}_e = 1.98 \times 10^{-8} \, W \, (K)^{-1} \, cm^{-3}$

$\dot{\lambda} = -5.91 \times 10^{-6} \, W \, cm^{-3}$

d) $\Delta \mathscr{E} = 2.58 \, V$

$\dot{s}_e = 1.98 \times 10^{-6} \, W \, (K)^{-1} \, cm^{-3}$

$\dot{\lambda} = -5.91 \times 10^{-4} \, W \, cm^{-3}$

IV.1. $R_{ij} = -\xi_{ij}$

IV.2. The conductance coefficients are generalizations of Eqs. (IV-20) and (IV-21) for *three* mobile components (1 = cation, 2 = anion, 3 = solvent, 4 = membrane, c_s = concentration of salt in membrane, mol salt cm^{-3} membrane volume. The membrane volume is the sum of solid and pore volume):

$$L_{11} = \frac{c_s^2}{c_w \Delta} [(f_{13} + f_{23})(f_{12} + f_{23} + f_{24}) - f_{23}^2] + \frac{c_s}{\Delta} [f_{34}(f_{12} + f_{23} + f_{24})]$$

$$L_{22} = \frac{c_s^2}{c_w \Delta} [(f_{13} + f_{23})(f_{12} + f_{13} + f_{14}) - f_{13}^2] + \frac{c_s}{\Delta} [f_{34}(f_{12} + f_{13} + f_{14})]$$

$$L_{33} = \frac{c_w}{\Delta} [(f_{12} + f_{13} + f_{14})(f_{12} + f_{23} + f_{24}) - f_{12}^2]$$

$$L_{12} = L_{21} = \frac{c_s^2}{c_w \Delta} [f_{12}(f_{13} + f_{23}) + f_{13}f_{23}] + \frac{c_s}{\Delta}(f_{12}f_{34})$$

$$L_{13} = L_{31} = \frac{c_s}{\Delta} [f_{13}(f_{12} + f_{23} + f_{24}) + f_{12}f_{23}]$$

$$L_{23} = L_{32} = \frac{c_s}{\Delta} [f_{23}(f_{12} + f_{13} + f_{14}) + f_{12}f_{13}]$$

where

$$\Delta \equiv \begin{vmatrix} (f_{12} + f_{13} + f_{14}) & -f_{12} & -f_{13} \\ -f_{12} & f_{12} + f_{23} + f_{24} & -f_{23} \\ \dfrac{-c_1 f_{13}}{c_3} & \dfrac{-c_1 f_{23}}{c_3} & \dfrac{c_1(f_{13} + f_{23})}{c_3} + f_{34} \end{vmatrix}$$

The units of the conductance coefficients, L_{ij}, are $W^{-1}\, mol^2\, cm^{-1}\, s^{-2}$.

IV.3. The electrokinetic equations are Eqs. (IV-44) and (IV-45). The conductance coefficients are $L_{vv} = 2.3 \times 10^{-4}\, cm\, s^{-1}\, MPa^{-1}$, $L_{ii} = 8\, \Omega^{-1}\, cm^{-2}$, $L_{iv} = L_{vi} = 3.75 \times 10^{-4}\, cm\, s^{-1}\, V^{-1}$. The streaming current is $3.76 \times 10^{-3}\, A\, cm^{-2}$.

V.1. Reciprocity [Eq. (V-5)] is satisfied, but since $L_{11}L_{22}(=2.13 \times 10^{-4}\, A^2)$ $< L_{12}^2(=156 \times 10^{-4}\, A^2)$, Eq. (V-4) is no satisfied. Hence the claim is invalid.

V.2. $q = L_{vi}/(L_{vv}L_{ii})^{1/2} = 0.00874$.

VI.1. $\Delta H = -67.2 \times 10^3\, cal\, (mol\, CO)^{-1} = -134.5 \times 10^3\, cal\, (mol\, O_2)^{-1}$

VI.2. All chemical potentials in $W\, s\, mol^{-1}$:
a) For iodine vapor of pressure p_{I_2} (atm):
   $\mu_{I_2, g} = 19389 + 2479\, ln\,(p_{I_2})$.
b) For aqueous solutions of iodine of molarity C_{I_2, H_2O} (mol l^{-1}), $\mu_{I_2, H_2O} = 16438 + 2479\, ln\,(C_{I_2, H_2O})$.
c) For solutions in carbon tetrachloride of molarity C_{I_2, CCl_4}
   $\mu_{I_2, CCl_4} = 5394 + 2479\, ln\,(C_{I_2, CCl_4})$.

VIII.1.
a) $-W = 0.701\, kWh\, tonne^{-1}$
b) $\Pi = 2.523\, MPa = 24.9\, atm$
c) 8872 m
d) The scheme would not work since the general potential, $\tilde{\mu}_W$, is uniform in either column. Hence the generalized driving force for the transfer of water between the columns, $-\Delta\tilde{\mu}_W$, does not vary with the depth.

VIII.2.
a) $L_{sw} = L_{ws} \simeq 0$.
b) Compare $H_2O \leftrightarrow THO$ self-diffusion through the membrane to osmosis under the action of the generalized driving force, F_w, on H_2O. The self-diffusion

mobility of water, m_d, is defined by Eq. (III-25), where $(-d\mu_w^c/dz) = F_w$ is the generalized driving force for self-diffusion of H_2O. The osmotic mobility, m_{osm}, is defined by the flux equation for osmosis:

$$J_w = m_{osm}\bar{c}_w F_w .$$

In terms of the friction model

$$m_d = 1/(f_{wm} + \xi_{ww*}c_{w,t}) ,$$

where $c_{w,t}$ is the total water concentration (H_2O + THO), and subscripts w and $w*$ designate H_2O and THO, respectively, and the osmotic mobility is

$$m_{osm} = 1/f_{wm} .$$

Hence

$$g_{do} \equiv m_{osm}/m_d = (f_{wm} + \xi_{ww*}c_{w,t})/f_{wm} \geq 1 .$$

c) Knudsen conditions in gas interdiffusion mean *uncoupling* of the flows of interdiffusing gases, due to the fact that the number of collisions between the two types of gas molecules is negligible compared to the number of collisions between gas molecules and walls. In this case, the hydraulic permeability can be calculated from interdiffusion measurements. When the flows of interdiffusing water molecules are *uncoupled* ($\xi_{ww*}c_{w,t}$ is negligible compared to f_{wm}), $g_{do} = 1$, i.e., the osmotic mobility can be calculated from self-diffusion measurements.

IX.1.

a) $p_{II} = 1.016$ atm.

b) The vessels at lower and higher pressure must be heated and cooled, respectively.

$$(\mathscr{J}_Q)_{\Delta T = 0} = -50.86 \text{ kcal min}^{-1} .$$

X.1.

a) To maintain the temperature uniform at 300°C, 9.00×10^{-3} cal h^{-1} is furnished at junction 2 and removed at junction 2'.

b) 2.58×10^{-3} cal h^{-1} is removed at junction 2 and furnished at 2' to keep the temperature uniform at 100°C.

c) 180.2°C.

d) At 300°C: -1.05×10^{-4} V (K)$^{-1}$.
   At 100°C: -0.297×10^{-4} V (K)$^{-1}$.

X.2.

a) $T_c = -\Delta T$.

Thermoelectric power

$$\equiv -\frac{d\Delta\mathscr{E}}{dT_c} = \frac{d(-\Delta\mathscr{E})}{d(-\Delta T)} = \frac{d(\Delta\mathscr{E})}{d(\Delta T)} = \eta_{SB} = A - B(\Delta T).$$

b) $(\eta_{SB})_{M_1-M_2} = (\eta_{SB})_{M_1-Pb} - (\eta_{SB})_{M_2-Pb} = (\eta_{SB})_{Pb-M_2} - (\eta_{SB})_{Pb-M_1}$.

c) $\Delta\mathscr{E}_{Pt-Ag} = 5.985\ T_c + 0.01963\ T_c^2\ \mu V$.

Subject Index

Wärme- und Stoffübertragung

Thermo- and Fluid Dynamics

Editors:

E.R.G. Eckert, University of Minnesota, Minneapolis, MN, USA

P. Grassmann, ETH Zürich, Zürich, Switzerland

U. Grigull, Technische Universität, München, FRG

F. Mayinger, Technische Universität, München, FRG

Problems of heat and mass transfer are being accorded increasing importance in industrial practice. Considerable efforts are being invested in research both experimental and theoretical – in most industrialized countries. **Wärme-und Stoffübertragung/Thermo- and Fluid Dynamics** is concerned with the latest research findings in this field.

Fields covered: Technical and Chemical Thermodynamics; Physical Chemistry; Chemical Process Engineering; Engineering Materials; Heating and Refrigeration; Nuclear Engineering; Rocketry and Reactor Technology; Boilermaking; Container, Piping and Turbine Construction.

Springer-Verlag
Berlin
Heidelberg
New York
Tokyo

Subscription information and/or **sample copies** are available from your bookseller or directly from Springer-Verlag, Journal Promotion Dept., P. O. Box 105 280, D-6900 Heidelberg, FRG

Experiments in Fluids

Experimental Methods and
Their Applications to Fluid Flow

Editors:
W. Merzkirch, Institut für Thermo- und Fluiddynamik,
Ruhr-Universität Bochum, Postfach 102148,
D-4630 Bochum, Federal Republic of Germany;
J.H. Whitelaw, Mechanical Engineering Department
Imperial College, Exhibition Road, London SW7 2BX,
Great Britain

Editorial Advisory Board:
T. Asanuma, Hiratsuka-shi Kanagawa-ken, Japan;
H.A. Becker, Kingston, Canada; **D. Bershader,** Stanford,
CA; **J.-M. Delhaye,** Grenoble, France; **F. Durst,** Erlangen,
FRG; **R.J. Emrich,** Bethlehem, PA; **R.J. Goldstein,**
Minneapolis, MN; **P. Hutchinson,** Harwell, GB; **D.G. Jones,**
Derby, GB; **S.J. Kline,** Stanford, CA;
M. Lapp, Schenectady, NY; **T.J. Mueller,** Notre Dame, IN;
E. Muschelknautz, Dormagen, FRG; **R.W. Nicholls,**
Downsview, Canada; **A.E. Perry,** Parkville, Australia;
R. Soloukhin, Minsk, USSR; **T. Tagori,** Tokyo, Japan;
Y. Tanida, Tokyo, Japan; **P.P. Wegener,** New Haven, CT;
F.J. Weinberg, London, GB; **W.-J. Yang,** Ann Arbor, MI.

Experiments in Fluids publishes research papers and technical notes which describe either:
- The development of new measuring techniques or the extension and improvement of existing methods for the measurement of flow properties necessary for the better understanding of fluid flows and their application in science and engineering.
- The application of experimental methods to the solution of problems of fluid flow.

The contributions will encompass a wide range of applications including aerodynamics, hydrodynamics, basic fluid dynamics, convective heat transfer, combustion, chemical, biological, and geophysical flows, and turbomachinery. Those which report advances to the analyses of flow problems will be considered provided they contain a substantial experimental content.

Springer-Verlag
Berlin
Heidelberg
New York
Tokyo

Subscription information and/or **sample copies** are available from your bookseller or directly from
Springer-Verlag, Journal Promotion Dept.,
P.O. Box 105280, D-6900 Heidelberg, FRG